Bioluminescence - Analytical Applications and Basic Biology

Edited by Hirobumi Suzuki

Published in London, United Kingdom

IntechOpen

Supporting open minds since 2005

Bioluminescence - Analytical Applications and Basic Biology
http://dx.doi.org/10.5772/intechopen.77783
Edited by Hirobumi Suzuki

Part of IntechOpen Book Series: Biochemistry, Volume 4
Book Series Editor: Miroslav Blumenberg

Contributors
Dan Close, Andrew Kirkpatrick, Tingting Xu, Steven Ripp, Gary Sayler, Hirobumi Suzuki, Yukimasa Kobayashi, Ikuko Sakai, Katsunori Ogoh, Tadasu Sano, Anvar Ismailov, Aleskerova Leyla, Alenina Kristina, Efremenko Elena, Ming-Luen Jeng, Anchana Thancharoen, Sirima Masoh, Hiroshi Ueda, Yuki Ohmuro-Matsuyama

First published in London, United Kingdom, 2019 by IntechOpen
IntechOpen is the global imprint of INTECHOPEN LIMITED, registered in England and Wales, registration number: 11086078, The Shard, 25th floor, 32 London Bridge Street
London, SE19SG - United Kingdom
Printed in Croatia

British Library Cataloguing-in-Publication Data
A catalogue record for this book is available from the British Library

Additional hard and PDF copies can be obtained from orders@intechopen.com

Bioluminescence - Analytical Applications and Basic Biology
Edited by Hirobumi Suzuki
p. cm.
Print ISBN 978-1-78984-784-0
Online ISBN 978-1-78984-785-7
eBook (PDF) ISBN 978-1-83962-196-3
ISSN 2632-0983

For EU product safety concerns:
IN TECH d.o.o., Prolaz Marije Krucifikse Kozulić 3, 51000 Rijeka, Croatia, info@intechopen.com or visit our website at intechopen.com.

IntechOpen Book Series
Biochemistry
Volume 4

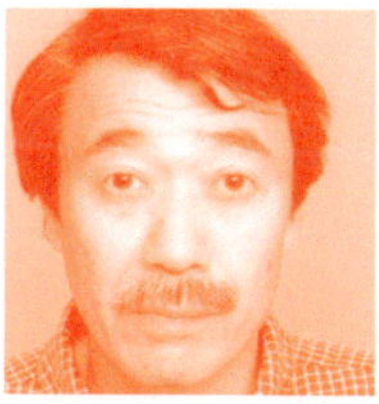

Dr. Hirobumi Suzuki received his PhD in Biology from Tokyo Metropolitan University, Japan, in 1997. His research interests include the phylogeny, speciation, and embryology of the firefly as well as bioluminescence technology. He is especially interested in the genetic differentiation pattern and speciation process in relation to flashing pattern as mating behavior of some fireflies in Japan. Dr. Suzuki currently serves as a fellow at the Olympus Corporation, a Japanese manufacturer of optics and reprography products. He is also a vice president of the Japan Fireflies Society. He and his colleagues developed a novel microscope for bioluminescence imaging that is currently being used for gene expression analysis in chronobiology, neurobiology, developmental biology, medical research, signal transduction analysis, molecular interaction analysis, and radiation biology.

Editor of Volume 4: Hirobumi Suzuki
Olympus Corporation, Tokyo, Japan

Book Series Editor: Miroslav Blumenberg
NYU Langone Medical Center, New York, USA

Scope of the Series

Biochemistry, the study of chemical transformations occurring within living organisms, impacts all of life sciences, from molecular crystallography and genetics, to ecology, medicine and population biology. Biochemistry studies macromolecules - proteins, nucleic acids, carbohydrates and lipids –their building blocks, structures, functions and interactions. Much of biochemistry is devoted to enzymes, proteins that catalyze chemical reactions, enzyme structures, mechanisms of action and their roles within cells. Biochemistry also studies small signaling molecules, coenzymes, inhibitors, vitamins and hormones, which play roles in the life process. Biochemical experimentation, besides coopting the methods of classical chemistry, e.g., chromatography, adopted new techniques, e.g., X-ray diffraction, electron microscopy, NMR, radioisotopes, and developed sophisticated microbial genetic tools, e.g., auxotroph mutants and their revertants, fermentation etc. More recently, biochemistry embraced the 'big data' omics systems.
Initial biochemical studies have been exclusively analytic: dissecting, purifying and examining individual components of a biological system; in exemplary words of Efraim Racker, (1913 –1991) "Don't waste clean thinking on dirty enzymes." Today

however, biochemistry is becoming more agglomerative and comprehensive, setting out to integrate and describe fully a particular biological system. The 'big data' metabolomics can define the complement of small molecules, e.g., in a soil or biofilm sample; proteomics can distinguish all the proteins comprising e.g., serum; metagenomics can identify all the genes in a complex environment e.g., bovine rumen. This Biochemistry Series will address both the current research on biomolecules, and the emerging trends with great promise.

Contents

Preface

A firefly is a bioluminescent organism that produces light via a chemical reaction inside its body. But what makes a firefly glow, exactly? To answer this question, one must consider both proximate and ultimate analyses of animal behavior. A proximate analysis looks at mechanistic and ontogenic explanations, while an ultimate analysis examines evolutionary explanations for adaptation and phylogenetic history. These aspects (mechanism, ontogeny, adaption, and phylogeny) are known as Tinbergen's Four Questions. Developed by Nikolaas Tinbergen, this method was originally designed to explain animal behavior. However, this categorization is also useful for other biological traits. Hence, it is used in this book to examine the trait of bioluminescence, which is increasingly being used in biomedical applications such as cellular assays and imaging.

Section 1 on "Bioluminescence and Its Applications" contains three chapters. The first chapter examines bioengineering advances that have been applied to firefly luciferase, *Gaussia* luciferase, *Renilla* luciferase, *Ophophorus* luciferase, and bacterial luciferase. The second chapter describes a bioluminescent system called NanoLuc, which uses luciferase from the sea shrimp *Oplophorus gracilirostris*. The final chapter in this section presents a biomonitoring system of toxins in an environment using photobacteria cells.

Section 2 on "Bioluminescent Organisms" also contains three chapters. The first chapter in this section reports on the ecology and histology of the light organ of a luminous springtail discovered in Japan. The second chapter examines the effect of artificial light produced by camera illumination on the flashing behavior of synchronous fireflies in Thailand. The final chapter reviews the biofluorescence of terrestrial organisms, planarians, snails, earthworms, nematodes, velvet worms, arthropods, and vertebrates. It also describes the phenomenon of both fluorescence and luminescence in fireflies using original field observations.

I would like to express my appreciation to the chapter authors for their outstanding work, and also to Ms. Ivana Barać, Author Service Manager at IntechOpen, for effective communication and assistance during the preparation of this book.

Hirobumi Suzuki
Olympus Corporation,
Tokyo, Japan

Section 1

Bioluminescence and Applications

Chapter 1

Biotechnological Advances in Luciferase Enzymes

Andrew Kirkpatrick, Tingting Xu, Steven Ripp, Gary Sayler and Dan Close

Abstract

This chapter explores the history of the bioengineering advances that have been applied to common luciferase enzymes and the improvements that have been accomplished by this work. The primary focus is placed on firefly luciferase (FLuc), *Gaussia* luciferase (GLuc), *Renilla* luciferase (RLuc), *Oplophorus* luciferase (OLuc; NanoLuc), and bacterial luciferase (Lux). Beginning with the cloning and exogenous expression of each enzyme, their step-wise modifications are presented and the new capabilities endowed by each incremental advancement are highlighted. Using the historical basis of this information, the chapter concludes with a prospective on the overall impact these advances have had on scientific research and provides an outlook on what capabilities future advances could unlock.

Keywords: firefly luciferase (FLuc), *Gaussia* luciferase (GLuc), *Renilla* luciferase (RLuc), *Oplophorus* luciferase (OLuc; NanoLuc), bacterial luciferase (Lux), biotechnology

1. Introduction

1.1 Historical perspective on the discovery of luciferase enzymes

The bioluminescent phenotype, which is spread across a variety of different insects, bacteria, fungi, and marine animals, has intrigued mankind since before the dawn of the modern scientific era [1]. The discovery that proteins, which would come to be known as luciferases, were responsible for bioluminescent production can be traced to early experiments by Raphael Dubois, who was able to produce bioluminescence *in situ* by mixing the contents of click beetle abdomens in cold water and extracting the components required for light production [2]. However, it was not until the late 1940s that the first luciferase protein was successfully purified from fireflies [3]. Around that same time, bacterial luciferase was elucidated and successfully expressed *in situ* [4]. However, despite the progress made with these luciferases, it would be some time until biotechnology had advanced to the point where the genes responsible for their expression could be cloned and exogenously expressed, setting off the use of luciferases as tools for scientific discovery [5, 6].

Following the exogenous expression of the previously described firefly and bacterial luciferases, *Renilla* luciferase was isolated from the sea pansy *Renilla reniformis* [7] and *Oplophorus* luciferase was isolated from the deep-sea shrimp, *Oplophorus gracilirostris* [8]. Shortly thereafter, firefly luciferase was successfully expressed in mammalian cells [9] and it was demonstrated that different luciferases could be used in tandem within a single host if they utilized different luciferin compounds [10]. More recently, *Gaussia* luciferase has been isolated from the marine copepod, *Gaussia princeps* [11], which was a notable discovery because, unlike alternative luciferases, it is naturally secreted and thus could be monitored without needing to sacrifice the host cell during luciferin treatment. Since the discovery of *Gaussia* luciferase there has been rapid development of these enzymes through genetic engineering, but little progress on the introduction of new systems. However, this was recently changed with the introduction of fungal luciferase as a novel luciferase system, which like bacterial luciferase is capable of genetically encoding both the luciferase and luciferin pathway genes to support autobioluminescent production [12].

1.2 Available luciferase systems for biotechnological applications

Of the ~40 different bioluminescent systems known to exist in nature [13], relatively few are available for biotechnological applications. The primary reasons for this are the lack of elucidated functional units, similarities in performance characteristics (such as wavelength output) relative to existing systems, the entrenchment of existing luciferase systems within the literature and as commercially-available products, and the relatively high monetary and time costs required to explore novel systems in depth relative to their ultimate utility as research tools. As a result of these barriers, the luciferases available as research tools are generally limited to those listed in **Table 1**.

1.3 The necessity of engineering luciferase proteins

Despite the variety of different luciferases available, it is impossible to identify just one that could fit the needs of every experimental design. Furthermore, it is unfortunately frequent that no luciferase can be found to fit the needs of a given experiment. As a result, there has been significant effort to engineer the existing luciferase enzymes to improve their functionality, make them easier to use, and expand their utility. This is especially true as the prevalence of luciferase usage has increased in biomedical applications, which rely upon human cellular and small animal model systems that have significantly different physical and biochemical properties relative to the native host organisms from which these proteins were sourced.

These changes in physical properties and the constraints applied by the needs of biomedical research have necessitated that luciferases be modified to express at longer output wavelengths that better penetrate animal tissues or that can be co-expressed with alternative luciferases, to produce light upon exposure to alterative luciferin compounds, to produce altered signal output kinetics that are shorter

Luciferase	Luciferin compound	Output wavelength (nm)
Firefly luciferase (FLuc)	D-luciferin	560
Bacterial luciferase (Lux)	Tetradecanal	490
Renilla luciferase (RLuc)	Coelenterazine	480
Oplophorus luciferase (OLuc)	Coelenterazine	460
Gaussia luciferase (GLuc)	Coelenterazine	470

Table 1.
Common luciferases available for biotechnological applications, their luciferin compound, and their output wavelength.

Technique	Common uses
Mutagenic PCR	Wavelength shifting, thermostability improvement, improve signal output intensity
Rational sequence mutation	Wavelength shifting, altering luciferin compatibility, altering signal output kinetics
Synthetic recapitulation	Enable functionality in alternative hosts, improve expression efficiency, improve ease of use
Codon optimization	Improve expression efficiency
Circular permutation	Thermostability improvement, improve expression efficiency, expand reporter functionality
Alternative luciferin supplementation	Wavelength shifting, altering signal output kinetics
Split luciferase complementation	Alter signal output kinetics, expand reporter functionality

Table 2.
Common approaches for engineering improvements in luciferase functionality.

or longer than their wild-type kinetics, to allow multimeric enzymatic structures to function as monomers, to stabilize or destabilize protein structure within the host, to make expression more efficient, and to increase output intensity so that it is easier to detect the signal. Imparting these changes makes it possible to utilize specialized versions of each luciferase that better fit the experimental needs of the researcher. As the breadth of luciferase usage continues to grow, and as new luciferase systems have been introduced over the years, the lessons learned from these modifications are refined and re-applied in order to continuously unlock new applications and improved functionality.

1.4 Common methods for engineering improvements

To support the need for continued luciferase improvement, a number of techniques have become commonplace for different engineering goals. The most commonly utilized approaches and their common engineering endpoints are shown in **Table 2**. Examples of the use of these techniques can be found in each of the following sections.

2. Firefly and click beetle luciferases

2.1 Background

Firefly luciferase (FLuc) is perhaps the most well-known, well-studied, and widely-used of all the luciferases. It, and its close relatives from click beetles, both function through the ATP-dependent oxidation of reduced D-luciferin (2-(4-hydroxybenzothiazol-2-yl)-2-thiazoline acid) in the presence of magnesium (Mg^{2+}) and molecular oxygen (O_2) to yield carbon dioxide (CO_2), AMP, inorganic pyrophosphate (PP_i), and oxyluciferin. The resulting oxyluciferin is initially produced in an excited state, and as it returns to its ground state energy is released in the form of light. The naturally occurring peak emission wavelength for FLuc (as commonly derived from *Photinus pyralis*) is ~560 nm, while click beetle luciferases, such as those from *Pyrophorus plagiophthalamus* and related species, can produce a variety of wavelengths from 537 to 613 nm depending on their source organism [14, 15].

Although FLuc and click beetle luciferase were among the first luciferases to be studied [16], it was not until the mid-1900s that significant progress was made in understanding the system at a level where it could be experimentally useful. At this time, McElroy successfully extracted firefly luciferase from purified firefly lanterns and determined that ATP was required for bioluminescence [17]. This led to the determination of D-luciferin's structure as 2-(4-hydroxybenzothiazol-2-yl)-2-thiazoline acid and its eventual chemical synthesis [16]. With these pieces in place, chemists were able to isolate oxyluciferin as a purified product of the luminescence reaction and validate its mechanism of action [18]. In 1985, FLuc cDNA was cloned by DeLuca et al. [19]. This provided an alternative to the use of crude extracts of beetles as a source of the luciferase enzyme and opened the door for widespread use in biotechnological applications.

2.2 Initial application and limitations

In its initial incarnation, FLuc was highly useful as a reporter in molecular biology and bioimaging studies and for assaying the presence and quantification of the metabolites that participate in or are connected to the light reaction. The early discovery that ATP concentration was proportional to light intensity in beetle luciferase reactions made this assay the primary method for monitoring the cell's main source of energy. Further entrenching this technology was its exceptional sensitivity. FLuc-based bioluminescent ATP assays display detection capabilities down to 10^{-17} mol [15]. This sensitivity for measuring ATP concentrations has been used in several applications including screening for microbial contamination in food industries, assessing cell viability [20], and assaying enzymes involving ATP generation or degradation [21]. However, ATP concentrations found in living cells (1–10 mM) are generally saturating for FLuc and therefore it cannot be routinely used to assay intracellular ATP content [15]. In a similar vein, FLuc has also been used to assay for the other metabolites that participate in its bioluminescence reaction: CoA, AMP, and PP_i [20].

The major limitation encountered during the use of FLuc or beetle luciferases has been the requirement that the luciferin substrate be exogenously provided for luminescence to occur. To date, there are no bacterial systems for generating luciferin *de novo*, which necessitates chemical synthesis and results in potential storage concerns due to the labile nature of the chemical [18]. Furthermore, this often requires that the host cell harboring the luciferase be lysed to enable substrate uptake, which has prevented its use for reporting real-time expression.

2.3 Engineering improved expression and output

Applications of wild-type beetle luciferases can be limited due to structural and functional stability issues or variations in the specific activity of the enzyme under varying temperatures, pHs, ion concentrations, or inhibitors [22]. For instance, wild-type FLuc protein has a half-life of only 15 minutes at 37°C. This required that more thermostable forms be developed to assay human and small animal model-relevant temperature conditions [23]. Pozzo et al. sought to address this issue by combining amino acid mutations shown to enhance thermostability with other mutations reported to enhance catalytic activity, resulting in an eight amino acid FLuc mutant that exhibited both improved thermostability and brighter luminescence at low luciferin concentrations [24].

Similarly, Fujii et al. produced variants capable of producing 10-fold higher luminescence than the wild-type enzyme by screening a mutant library of FLuc proteins generated by random mutagenesis [25]. Site-directed mutagenesis experiments were then performed based on mutant sequences that produced increased luminescence. It was observed that the substitution of D436 with a non-bulky amino acid, I423 with a hydrophobic amino acid, and L530 with a positively charged amino acid all increased luminescence intensities relative to the wild-type enzyme. They further demonstrated that combining the mutations at I423, D436, and L530 resulted in an overall increase in affinity and turnover rate for the ATP and D-luciferin substrates that resulted in high amplification of luminescence intensity. Studies like this represent an emerging trend of combining alterations to specific properties of firefly luciferases in order to enhance its overall practical utility.

2.4 Engineering alternative output wavelengths

Engineering wavelength-shifted luciferases has become an intense area of study to enable multi-color assays and improve the efficiency of *in vivo* bioimaging. Due to hemoglobin's absorbance of wavelengths below 600 nm in mammalian tissues, the use of wild-type firefly luciferase is relatively handicapped compared to more red-shifted variants [15]. To overcome this limitation, mutagenic engineering approaches have been successfully used to generate a variety of red-shifted versions [26, 27]. Notable among this group is a variant developed by Branchini et al. containing a S284T mutation. This variant produces a red-shifted output with a peak at 615 nm, a narrow emission bandwidth, and improved kinetic properties [26]. However, this is by no means the only option available. Today, the wide variety of available output wavelengths enables researchers to choose the variant most well suited to their needs, or multiple variants that can be simultaneously triggered upon exposure to D-luciferin.

2.5 Engineering alternative signal kinetics

It has been demonstrated that varying the concentrations of FLuc's substrates (D-luciferin, ATP, etc.) can alter its reaction kinetics. High or saturating concentrations produce flash-type kinetics that result in an intense initial signal followed by a rapid decay, while low concentrations produce glow-type kinetics with a relatively lower initial signal and a slower decay [18]. There are many possible inhibitors that could be responsible for these changes. Under high substrate conditions, byproducts of the reaction such as oxyluciferin and L-AMP can act as tight active-site binding inhibitors preventing enzyme turnover, or inhibitor-based stabilization can increase activity when substrate levels are high enough to compete with the inhibitory compound [14]. Commercial reagents containing micromolar concentrations of components such as pyrophosphate and/or CoASH have been shown to convert FLuc reactions from flash- to glow-type kinetics, possibly due to the breakdown of oxidized luciferin-AMP *via* pyrophosphorolysis and thiolysis into the less potent inhibitors oxidized luciferin and oxidized luciferin-CoA, respectively. These commercial reagents are now widely used to support different experimental needs [14].

Another strategy that has been applied to alter reaction kinetics is the modification of the luciferin substrate. Mofford et al. demonstrated that near-infrared light emission can be increased >10-fold from wild-type FLuc by replacing D-luciferin with synthetic analogues [28]. These synthetic analogues were designed to emit longer wavelength light by incorporating an aminoluciferin scaffold. Nearly all the aminoluciferins tested in their studies resulted in higher total near-IR (695–770 nm) photon flux from live cells under both high- and low-dose conditions. A more recent substrate modification strategy has been to conjugate the luciferin with distinctive functional groups. These so-called "caged" luciferins react when they are cleaved by enzymes or bioactive molecules and subsequently freed [29]. This strategy allows for specified monitoring of biological processes by linking light output to the activity and/or concentration of enzymes or molecules reacting to cleave the caged luciferins.

3. *Renilla* luciferase

3.1 Background

Like FLuc, *Renilla* luciferase (RLuc) is another commonly used bioluminescent reporter. Derived from the sea pansy *Renilla reniformis*, RLuc is a decarboxylating oxidoreductase that uses coelenterazine as its substrate. During its bioluminescent reaction, coelenterazine is converted to coelenteramide in the presence of molecular oxygen, yielding blue light with an emission peak at 480 nm [30]. In addition to their substrate preferences, one other important differentiator between RLuc and FLuc is that the RLuc bioluminescent reaction does not require ATP. It is also significantly less efficient than FLuc and produces a reduced relative light output intensity with a quantum yield of ~7% [31].

The RLuc protein was first purified and characterized in the late 1970s [7]. However, its cDNA sequence was not identified and cloned into *Escherichia coli* until 1991 [31]. Following that accomplishment, the recombinant RLuc protein was quickly expressed in other organisms, including yeast [32], plant [33], and mammalian cells [34] to serve as a gene expression reporter. The successful detection of RLuc bioluminescence from mammalian cells was particularly important because it represented the proof-of-principle demonstration of this enzyme as a reporter target for *in vivo* animal imaging. And indeed, imaging of RLuc activity in living mice was successfully validated just several years later [35]. In this demonstration, Bhaumik and Gambhir showed that intraperitoneally implanted RLuc-expressing cells could be detected following the injection of coelenterazine into the tail-vein [35]. Similarly, when cells were injected *via* the tail-vein, bioluminescent signal could be used to visualize cell trafficking to the liver and lungs. This study also validated that D-luciferin could not be used as a substrate, opening the door for future studies to multiplex RLuc with FLuc as dual-reporters for *in vivo* applications.

3.2 Engineering improved expression and output

The initial limitation for using RLuc as a reporter was its less-than-optimal expression efficiency within mammalian cellular hosts. This limitation was overcome *via* a codon-optimization strategy that modified the RLuc gene sequence while maintaining the wild-type protein sequence. A synthetic humanized version of the luciferase gene that utilizes this strategy, called hRLuc, is now commercially available and has been shown to produce up to several 100-fold higher light output in many mammalian cell lines. Further hampering the expression of RLuc in cell culture and small animal imaging applications was its tendency to be rapidly inactivated upon exposure to animal serum. In its wild-type orientation the half-life of the enzyme under routine experimental conditions ranged from 30 to 60 minutes [36]. An early study by Liu and Escher showed that a single mutation from cysteine to alanine at amino acid 124 (RLucC124A) increased serum resistance, while simultaneously increasing overall light output [37]. Following this study, Loening et al. employed a consensus sequence guided mutagenesis strategy to screen for mutants with improved serum stability [36]. These efforts identified a variant termed RLuc8, which harbored eight substations (A55T/C124A/S130A/L136R/A143M/M185V/M253L/S287L). The RLuc8 variant was shown to be >200-fold more stable in mouse serum than the native protein and displayed an improved half-life of 281 hours. Fortuitously, the RLuc8 mutant also exhibited a 4-fold improvement in brightness. The improved stability and light output characteristics of RLuc8 make it a more favorable reporter than wild-type RLuc for mammalian imaging applications.

3.3 Engineering alternative output wavelengths

Despite the improvements made to increase expression efficiency and output, RLuc's 480 nm output maximum remained problematic for *in vivo* animal imaging applications because it was prone to absorption and attenuation in organs and tissues. This was especially problematic for deep tissue imaging (below subcutaneous layer), where only 3% of the emission spectra could efficiently penetrate animal tissue for detection. Therefore, to improve RLuc's *in vivo* utility, many efforts were undertaken to red-shift its emission spectra. Loening et al. hypothesized that modifying the active site of the luciferase could create a chemical environment favorable to specific coelenteramide species (i.e., the pyrazine anion form) that emit green (535–550 nm) light upon returning from their excited state. To test this hypothesis, they made site-specific mutations at 22 amino acid residues at the predicted active site of RLuc8 and identified red-shifted light emissions (peaks between 493 and 513 nm) in variants with mutations at eight of these locations [38]. Unfortunately, these red-shifted mutants also possessed substantially reduced signal intensities. To restore light output, random mutagenesis was carried out on the red-shifted mutants. This process identified several residues where mutations increased light output or resulted in further red-shifting. Based on these encouraging results, Loening and colleagues performed several more rounds of site-directed mutagenesis and successfully engineered three promising variants RLuc8.6-535, RLuc.6-545, and RLuc8.6-547, which peaked at 535, 545, and 547 nm, respectively, when using coelenterazine as the substrate. All three variants exhibited greater light output than wild-type RLuc, with the most improved, RLuc8.6-535, showing 6-times greater intensity and similar stability to RLuc8. In practice, this translated to roughly a 2.2-fold increase in transmitted signal from the lungs of living mice compared to an equal initial light flux from RLuc8.

In addition to engineering the protein itself, synthetic coelenterazine substrate analogs have also been created to improve light output and/or yield red-shifted emission spectra. The analog coelenterazine-*v* was first shown to shift the emission peak of wild-type RLuc to 513 nm [39] and later demonstrated to yield emission peaks at 570 nm (yellow) and 588 nm (orange) in the RLuc8.6-535 and RLuc8.6-547 variants, respectively [38]. However, this substrate is currently not commercially available due to high background activity and difficulty in purification. Other analogs, such as coelenterazine-*f*, -*h*, and -*e* have been shown to increase signal intensity by 4- to 8-fold relative to coelenterazine in RLuc-expressing mammalian cells *in vitro*, but each has failed to compete with the native coelenterazine in living animal imaging [40]. Despite these setbacks, Nishihara et al. have reported that analogs with ethynyl or styryl group substitutions at the C-6 position significantly increased bioluminescent output and signal stability in RLuc8 and RLuc8.6-535 [41, 42], which suggest that the development of new synthetic coelenterazine analogs will continue to be a promising route for enhancing RLuc functionality.

3.4 Engineering split luciferase applications

Due to its small size (311 amino acids, ~36 kDa) and monomeric orientation, the RLuc protein is an attractive option for use in split luciferase complementation assays aimed at monitoring real-time protein-protein interaction. In an early study attempting to achieve this goal, Paulmurugan and Gambhir [43] created RLuc fragment pairs at two split sites (I223/P224 and G229/K230) and fused the individual fragments to either the MyoD or Id proteins. They then successfully demonstrated that RLuc could properly re-fold and restore luciferase activity upon complementation during MyoD/Id interaction. This study also showed that the split RLuc reporter signal could be modulated by using an inducible promoter (e.g., NFκB promoter/enhancer) to regulate the expression level of one of the two fragments. The fragment pair based on the G229/K230 split site was later used to characterize interactions between heat shock protein 90 (Hsp90) and the co-chaperone protein Cdc37 [44], between Hsp90 and the Epstein-Barr virus protein kinase GBLF4 [45], and to visualize androgen receptor translocation in the brains of living mice [46]. Kaihara et al. similarly leveraged a variant of RLuc split between S91 and Y92 to demonstrate the recovery of bioluminescent activity during insulin-stimulated protein-protein interactions [47], and Stefen et al. created a split variant using fragments separated between residues 110 and 111 fused to protein kinase A (PKA) regulatory and catalytic subunits to quantify G protein-coupled receptor (GPCR)-induced disassembly of the PKA complex in living cells [48]. These types of split RLuc complementation assays have also been applied to profile protein-protein interactions in the Golgi apparatus *in planta* [49] and to study protein dynamics during chemotaxis in bacteria [50], making it a broadly applicable approach.

4. *Gaussia* luciferase

4.1 Background

Isolated from the marine copepod *Gaussia princeps*, *Gaussia* luciferase (GLuc) is the smallest known luciferase. It is comprised of only 185 amino acids and has a molecular weight of 19.9 kDa. Like RLuc, GLuc catalyzes the oxidative decarboxylation of coelenterazine in an ATP-independent manner to produce blue light with a peak wavelength around 480 nm. Despite this relatively short wavelength, GLuc is one of the brightest luciferases and is capable of generating light output several orders of magnitude higher than FLuc and RLuc [11]. However, unlike FLuc and RLuc, the GLuc protein is naturally secreted from the cells. In biotechnological applications, this allows signal measurements to be performed on culture medium without cell lysis and when using blood or urine samples obtained during animal applications [51, 52]. Its secretory nature also enables unique applications such as monitoring protein processing through the secretory pathway and drug-induced endoplasmic reticulum (ER) stress [53, 54]. It was first isolated and cloned by Bruce and Szent-Gyorgyi in 2001 [55], and since has enjoyed rapid adoption within the research community through a variety of engineered improvements.

4.2 Engineering improved expression and output

To enable improved expression efficiency in biomedical applications, a humanized version of GLuc, hGLuc, was generated *via* codon optimization. This human-optimized variant has been shown to produce 2000-fold higher bioluminescent signal than the wild-type variant when expressed in mammalian cells [11]. In addition to mammalian systems, the GLuc gene sequence has also been codon optimized for efficient expression in the alga *Chlamydomonas reinhardtii* [56], the fungus *Candida albicans* [57], mycobacteria [58], and *Salmonella enterica* [59].

Building on this codon optimization-based approach, which enhances light output by improving protein expression in the host organism without modifying the peptide sequence, mutagenetic approaches have similarly been successfully applied to engineer variants that produce greater signal intensities than the wild-type protein. In one such example, Kim et al. performed site-directed mutagenesis to the hydrophilic core region of GLuc and identified that changing the isoleucine at position 90 to leucine (I90L) was the major contributing factor for improved signal intensity [60]. The I90L variant produced six times higher light output than the wild-type protein in mammalian cells. Using a directed molecular evolution approach, Degeling et al. also identified a variant (S16K/M43V/V159M) that showed 2-fold enhanced luciferase activity [61].

4.3 Engineering alternative output wavelengths

One limitation of the native GLuc protein is that its relatively blue-shifted emission wavelength is easily absorbed and scattered by pigmented molecules in animal tissues. This limits its utility in *in vivo* animal imaging applications. Several attempts have been made to engineer a red-shift towards increased wavelengths, but these efforts have met with only moderate success. In one notable example, Kim et al. engineered a variant, which they termed Monsta, that harbors four mutations (F89W/I90L/H95E/Y97W) resulting in a shifted peak emission wavelength of 503 nm. This is ~20 nm red-shifted compared to wild-type GLuc [60]. Similarly, several alternative variants (L40P, L40S, and L30S/L40P/M43V) generated by Degeling et al. show 10–15 nm shifts in their emission peaks [61]. Despite the fact that these red-shifted variants have not enjoyed similar success to those of RLuc, GLuc's relatively increased signal strength can often compensate for the loss of signal due to absorption.

4.4 Engineering alternative signal kinetics

Wild-type GLuc catalyzes a flash-type bioluminescent reaction, meaning that the light signal decays rapidly following luciferin exposure. Practically, this necessitates immediate signal reading after substrate addition and thus makes GLuc unsuitable for the majority of high-throughput applications. To overcome this rapid signal decay, researchers have successfully engineered mutants that emit more stable bioluminescence [61–63]. Noticeably, a L30S/L40P/M43V variant has been shown to exhibit glow-type kinetics with only a 20% loss in signal intensity over 10 minutes, compared to the >90% loss in signal intensity after 1 minute from the wild-type enzyme [61]. GLuc mutants such as these have been demonstrated to function in 96- and 384-well plate formats, which effectively allows them to overcome the wild-type kinetic limitations and enables their use in high-throughput assay formats.

4.5 Engineering split luciferase applications

Like RLuc, GLuc's small size (185 amino acids, 19.9 kDa) makes it a good candidate for split luciferase complementation assays. In an early attempt at developing this functionality, Remy and Michnick evaluated the ability of fragment pairs generated from cut sites between amino acids 65–109 of a truncated hGLuc sequence exclusive of the secretion signal to reconstitute luciferase activity upon rejoining [64]. By fusing the respective 5′ and 3′ sequences of the split hGLuc gene to a GCN4 leucine zipper-coding sequence and co-expressing the resulting fusions in HEK293 cells they were able to show that hGLuc activity could be successfully reconstituted by leucine zipper-induced complementation of the split fragments. Their study determined that the optimal split site for complementation was between G93 and E94. This fragment pair has since been further demonstrated to be inducible and reversible, which allows it to function as a highly sensitive tool for quantifying protein-protein interactions in cells and living mice [65]. Similarly, Kim and colleagues also developed a split GLuc variant dissected at Q105 and demonstrated its utility to monitor calcium-induced calmodulin and M13 peptide interaction, phosphorylation of the estrogen receptor, and steroid-receptor binding in living cells [60].

5. *Oplophorus* luciferase

5.1 Background

Oplophorus luciferase (OLuc) is a naturally-secreted luciferase isolated from the decapod *Oplophorus gracilorostris*, a deep-sea shrimp that ejects OLuc from the base of an antennae in a brightly luminous cloud when stimulated. It is one of the more complex luciferase proteins, as it is a 106 kDa heterodimeric tetramer consisting of two regions, each comprised of a 35 and 19 kDa subunit. Like RLuc and GLuc, OLuc uses coelenterazine as a substrate and does not require ATP for functionality [17]. It produces primarily blue light, with a peak emission wavelength of 462 nm. Even in its wild-type form, OLuc possesses robust biochemical and physical characteristics relative to alternative luciferases. It exhibits relatively little change in quantum yield throughout a pH range from 6 to 10, maintains thermostability across a temperature range of 20–50°C, and can still produce observable light output at 70°C [8].

OLuc was first discovered in 1975 [66], and shortly after in 1978 the mechanics of its bioluminescent reaction were identified [8]. Inouye et al. were the first to clone the OLuc cDNAs encoding the 35 and 19 kDa subunit proteins, which led to their discovery that the 19 kDa protein was responsible for catalyzing the luminescent oxidation of coelenterazine. Although this 19 kDa protein was found to be the smallest known protein capable of catalyzing bioluminescence, it was also found to be poorly expressed and unstable without the support of its 35 kDa partner [67].

5.2 Engineering improved expression and output

The need to co-express the 19 and 35 kDa subunits of OLuc made it problematic for routine reporter usage. To overcome this, Hall et al. performed three rounds of mutagenesis on the 19 kDa subunit to produce a novel variant, which they termed NanoLuc (NLuc). This variant showed improved structural stability as well as increased bioluminescent activity and glow-type kinetics with a peak emission wavelength of 460 nm. Furthermore, it was shown that this variant could oxidize an alternative luciferin, furimazine, which resulted in greater light intensity and lower background autoluminescence than when coelenterazine was used. NLuc's 19 kDa size and absence of post-translational modifications made it more agile than FLuc, while its naturally high tolerance to temperature and pH made it more robust. In practice, this NLuc variant was shown to poses 150-fold greater specific activity than either FLuc or RLuc [68]. However, these improvements proved to be a double-edged sword. The high stability and glow-type kinetics made it difficult to employ NLuc for transient reporting activities, while its highly blue-shifted output limited its signal penetration in mammalian cellular applications.

Nonetheless, NLuc's small size and efficient expression make it an excellent choice for studying low-dynamic activities. In one such example, Chen et al. developed a sensitive assay in which NLuc was used to study the activity of deubiquitinating enzymes. In this work, NLuc was fused to the C-terminus of His-tagged ubiquitin that was attached to Ni^{2+} agarose beads. This allowed NLuc to be released as the α-peptide linkages were cleaved so that deubiquitination could be monitored *via* NLuc luminescence [69]. Similarly, Lackner et al. [70] used a CRISPR-Cas9-mediated strategy to tag three cytokine-inducible genes (DACT1, IFIT1, and EGR1) with NLuc. This allowed cytokine-induced upregulation to be measured in HAP1 cells. Under this design, they were able to show that NLuc luminescence correlated strongly with quantitative PCR data, demonstrating that NLuc could reliably be used to monitor gene expression.

5.3 Engineering split and paired luciferase applications

Zhao et al. showed that a split luciferase-based system could be used to monitor protein stability by tracking protein aggregation with NLuc-based luminescence [71]. To accomplish this, they broke NLuc into two fragments, termed N65 and 66C, and demonstrated that, upon interaction, luminescence was modulated by the solubility of the protein fused to the N65 fragment. This property was maintained in both bacterial and mammalian systems, confirming its utility for sensitive detection of protein solubility in a straightforward, high-throughput assay format in living cells.

In addition to these traditional split luciferase applications, NLuc has also been employed for paired luciferase applications that utilize an unfused variant to provide the highest possible light intensity and sensitivity, a destabilized variant with an appended degradation signal (e.g., NLuc-PEST) that allows rapid response to dynamic changes in environment, and a secreted variant (e.g., secNLuc) [17].

6. Bacterial luciferase

6.1 Background

Unlike the monomeric luciferases discussed above, bacterial luciferase (Lux) is a heterodimer of two genes, *luxA* and *luxB*, that must join together to form a functional unit. It is also only one of two systems, along with the fungal system discussed below, that additionally has a known genetic pathway for luciferin synthesis. In the case of bacterial luciferase, this pathway consists of three additional genes, *luxC*, *luxD*, and *luxE*, that work together to produce a long chain fatty aldehyde [72]. In this process, *luxD* transfers an activated fatty acyl group to water, forming a fatty acid. The fatty acid is then passed off to *luxC* and activated *via* the attachment of AMP to create a fatty acyl-AMP. The *luxE* gene finally reduces this fatty acyl-AMP to an aldehyde [72]. The natural aldehyde for this reaction is tetradecanal, however, the luciferase is also capable of functioning with alternative aldehydes as substrates [72]. Along with these genetic components, the system requires two cofactors: oxygen and reduced riboflavin phosphate. When all components of the system are present, bacterial luciferase will produce bioluminescence in an autonomous fashion at a wavelength of 490 nm.

Although this process has been most well-studied in marine bacteria from the *Vibrio* genus, the genetic organization and biochemical underpinnings of the system are consistent across all known bacterial phyla [18]. Due to the complexity of this system relative to its monomeric counterparts, it was not exogenously expressed until the early 1980s. Even then, it was initially utilized through expression of the *luxA* and *luxB* genes as a standalone luciferase [5] before subsequently being employed as a fully functional cassette that was capable of functioning in an autonomous fashion [73]. Shortly after these demonstrations the crystal structure of the bacterial luciferase heterodimer was determined [74], however, this structural knowledge has yet to be leveraged as a means for engineering improved functionality.

6.2 Initial uses and limitations

Because Lux emits its bioluminescent signal without the need for external stimulation, it quickly became a valuable tool for optical imaging. The low hanging fruit for this system was the real-time monitoring of gene expression. This was first demonstrated by Enbreghet et al. [75], who fused Lux to inducible promoters to study the mechanics of IPTG and arabinose induction in *E. coli*. This proved to be a valuable approach because it allowed samples to be continuously monitored in order to track gene expression dynamics over time. Building upon this work, a variety of instances have been described where Lux has been placed under the control of a promoter with a known inducer to track compound bioavailability. Repeated use of the system for this propose has demonstrated that it is capable of reporting bioavailability in a dose/response fashion [76], which makes it a valuable tool for monitoring contaminant levels in mixed environmental samples. At a higher level, it has been used for *in situ* bacterial monitoring, such as the visualization of bacterial invasion of leaf [77] and root structures [78]. Further, due to the absence of light production from non-bioluminescent species, it was also used to track specific populations of bacteria within mixed communities within unperturbed environments [79].

Despite the advantages offered by avoiding the need for external stimulation concurrent with visualization, Lux was significantly handicapped by its inability to function within eukaryotic cells. Because of this, it was not originally applicable to most modern biotechnological and biomedical applications outside of tracking bacterial infections [80]. Furthermore, as a consequence of encoding both the luciferase and luciferin generation pathways this system required significantly more foreign DNA to be introduced in order to function exogenously. This made the system more difficult to work with at the molecular level; especially before the advent of today's more efficient genetic assembly tools. Similarly, the heterodimeric nature of the luciferase enzyme is more cumbersome than the monomeric orientation of its counterparts. Nonetheless, given its relative advantages over the other systems, it continues to be engineered to overcome these detriments and expand its utility.

6.3 Engineering eukaryotic expression

Although several early attempts were made to enable Lux functionality within eukaryotic hosts, none of these achieved significant success [81–83]. The first major breakthrough came with the expression of the luciferase in *S. cerevisiae* [84]. This achievement was made possible by using luciferase genes from the terrestrial bacterium, *Photorhabdus luminescens*, which showed higher thermal stability than those of marine bacteria, and expressing the individual heterodimer genes from a single promoter using an internal ribosomal entry site (IRES) to link them together. Under this orientation the luciferase was able to properly express within the cell and produce light upon exposure to an n-decanal substrate. This same strategy was then expanded to incorporate the expression of IRES-linked luciferin synthesis pathway genes from dual promoters. When expressed concurrently with the luciferase genes, the cell produced a bioluminescent signal without external stimulation. The functionality of the system was then further improved by shifting the intracellular redox balance to a more reduced state through the introduction of a flavin oxidoreductase gene, *frp*.

Despite this success in *S. cerevisiae*, the direct application of these changes was not sufficient to permit similar bioluminescent production from human cellular hosts. To achieve this, the genes were codon optimized for the human genome and mammalian-optimized IRES elements were employed to improve expression of the downstream genes in human cells [85]. It was also determined that the full pathway could not be expressed from a single promoter using IRES elements, so the luciferin synthesis pathway was encoded on a separate plasmid. This approached allowed for functionality in human cells, but the overall level of bioluminescent production was several orders of magnitude lower than that of alterative bioluminescent systems such as firefly luciferase [86].

6.4 Engineering increased light output

To overcome Lux's low level of bioluminescent output in human cells the orientation of the cassette was subjected to further engineering. It was determined that the use of multiple plasmids was detrimental to achieving high level expression, and that the use of IRES elements was inefficiently expressing the downstream genes in the paired orientation. Therefore, the IRES elements were replaced with viral 2A linker sequences. These sequences were significantly shorter than the IRES sequences they replaced and allowed for each linker region to have a unique genetic code that reduced the chance for unintended recombination events. As a result, the full bacterial luciferase cassette, inclusive of the flavin oxidoreductase component, could be placed under the control of a single promoter and expressed from a single plasmid. This new orientation made it possible to express bacterial luciferase as a single genetic construct similar to what was commonly done with the alternative monomeric, luciferin-requiring luciferase systems. As a result, the bacterial luciferase system could be expressed more easily across a larger number of cell types and was capable of producing an enhanced level of signal output relative to its previous incarnation [87].

In addition to engineering increased expression *via* improved expression efficiency, work has also been performed to alter the peptide sequence of the bacterial luciferase genes to make light output more efficient. Gregor et al. [88] used random mutation to alter the coding sequence of Lux cassette and uncovered a series of 15 mutations that improved light output and thermostability. Of these mutations, six were within the luciferase genes (three each in *luxA* and *luxB*), six were in the luciferin synthesis pathway (with all six located in the *luxC* gene), and three were located in the oxidoreductase gene, *frp*. These mutations resulted in both improved thermotolerance and a ~7 times increase in bioluminescent production relative to the wild-type sequence.

6.5 Engineering improved bioreporter functionality

Just as it has been used extensively as a bioreporter in bacterial species, the engineering of bacterial luciferase to function in eukaryotic cells opened the door to this same functionality under much broader applications. The transition of the Lux cassette to function as a single open reading frame made it possible to replace the constitutive promoter with an inducible promoter and regulate its expression in response to compound bioavailability [87]. However, computational modeling aimed at calculating the metabolism of the required substrates and cofactors for the reaction relative to their intracellular availably suggested that that control of the system should be imparted at the level of the aldehyde recycling pathway, with *luxA* and *luxB* expressed continuously, and *luxC*, *luxD*, and *luxE* placed under the control of the inducible promoter [89]. This model was later proven to be correct when direct comparisons were performed using either single open reading frame constructs where the full cassette was controlled by the inducible promoter, or split cassettes where the luciferase and luciferin pathway genes were switched between inducible and constitutive promoters [90]. Together, these results significantly improved the functionality of the bacterial luciferase system as a bioreporter despite its relative complexity compared the other luciferases.

7. Fungal luciferase

7.1 Background

Fungal luciferase is the most recent luciferase system to be functionally elucidated and made available for biotechnological applications. At the core of this system is a monomeric luciferase gene, *luz*. In addition to the luciferase, two luciferin synthesis genes: *hisps* and *h3h*, work together as a polyketide synthase and a 3-hydroxybenzoate 6-monooxygenase to supply the required luciferin, 3-hydroxyhispidin. In addition to these genetic components, the reaction also requires molecular oxygen and NAD(P)H as co-factors [91, 92]. When all components of the system are present, it produces a luminescent signal at 520 nm. Like Lux, fungal luciferase is notable in that the genetic sequence of all components required for bioluminescent production is characterized. This allows the fungal luciferase cassette to be genetically encoded and exogenously expressed to produce an autobioluminescent phenotype [12]. However, for this to occur the host organism must either be capable of naturally synthesizing caffeic acid to act as a precursor for luciferin synthesis, or the necessary genes for caffeic acid synthesis must be co-expressed. Under this strategy, it is possible to synthetically assemble a seven gene cassette consisting of the fungal luciferase genes: *luz*, *hisps*, and *h3h*, along with a tyrosine ammonia lyase, two 4-hydroxyphenylacetate 3-monooxygenase components and the 4′-phosphopantetheinyl transferase gene *npgA*, to support caffeic acid synthesis and continuous light production in any host.

7.2 Initial uses and limitations

Unlike the previous luciferases that have been discussed, fungal luciferase has only recently been elucidated as of the time of this chapter. As a result, there have yet to be any reports of its functionality outside of its initial validation [12]. Regardless, the initial characterization of the system provides valuable insights into its functionality and potential limitations. From a practical standpoint, it has been demonstrated that the system can be fully recapitulated in yeast to achieve autobioluminescent signal production. At this time only one luciferin synthesis pathway has been demonstrated, but because genes sourced from alternative organisms are used to enable caffeic acid synthesis in hosts that do not natively support these reactions, it is likely that alterative genes could be substituted for these parts of the pathway.

For more complex hosts, such as human cells, the functionality of the system has been demonstrated only under non-autobioluminescent conditions. In this case, only the luciferase was genetically encoded and the luciferin was exogenously applied. Using this strategy, it has been possible to observe luminescence in cultured human cells, *Xenopus laevis* embryos, and small animal models subcutaneously injected with labeled cells. These demonstrations bode well for the use of the fungal luciferase in the types of experimental designs most commonly associated with traditional luciferase reporters and provide researchers with a novel imaging tool that can be differentiated from alternative luciferases based on its luciferin specificity.

It is currently unknown if the lack of demonstrated autobioluminescent production in hosts outside of yeast is incidental, or if it is the result of metabolic or molecular limitations on the expression of the full cassette within these organisms. One possible explanation is that the required culture temperatures were not compatible with full cassette functionality. It has been shown that fungal luciferase is temperature sensitive and begins to decrease its output signal at temperatures >18°C. Relative light output is halved at room temperature (26°C) and is abolished above 30°C. This is detrimental to the use of this luciferase in human cell culture and small animal model systems, as they will require the maintenance of temperatures above 30°C to avoid the introduction of secondary environmental effects. Similarly, the luciferase is only ~50% efficient at pH 7, which could be detrimental to some experimental designs. The optimal pH is 8, with improved retention of performance at increased pH relative to decreased pH.

7.3 Potential future engineering goals

There are ~100 fungal species that use this luciferase/luciferin pathway for bioluminescent production [93]. It is believed that fungal bioluminescence evolved only once, but that evolutionary pressure led to uneven distribution of the phenotype among species. While this simplifies the system by allowing development to focus on only a single incarnation, it is also potentially limiting in that there are fewer evolutionary cues that can be leveraged as starting points for biotechnological advancement. Nonetheless, this system is clearly in its infancy and will benefit from the copious knowledgebase developed through the engineering of alternative luciferases. It is likely that the primary development target will be overcoming the thermostability issues present in the current incarnation of the system. Beyond this, and similar to Lux, it is likely that investigators will seek to streamline expression of the relatively large cassette size to make it more manageable from a molecular biology standpoint. Once these efforts are achieved, the autobioluminescent nature and somewhat red-shifted output of the fungal system will make it a welcome addition for real-time imaging applications that currently rely on only the bacterial luciferase system.

8. Outlook for future developments

There are ~40 different bioluminescent systems known to exist in nature [13]. However, only seven different families have been well described and only five of the six detailed in this chapter enjoy widespread use [94]. Despite the relative wealth of unexplored systems, relatively few new systems have become available in recent history. Within the last 10 years, the most notable advancements have been the engineering of the bacterial luciferase system to function in eukaryotic organisms and the elucidation of the fungal luciferase genetic pathway. Despite this, the considerable progress of incremental engineering for firefly luciferase and the development of NanoLuc from *Oplophorus* luciferase have provided a clear roadmap for continued progress within the field. Historically, the ability to alter luciferase conformation or luciferin compatibility to enable alternative output wavelengths that better penetrate tissue, allow for multiplexed imaging of multiple luciferases, or pair with fluorescent reporters for BRET applications has enabled new experimental designs that have led to important discoveries. With the emergence of autobioluminescent capabilities from the bacterial and fungal systems, it is likely that the barriers will again be pushed back. These systems will compete with the established luciferases and encourage further development to keep them competitive within an increasingly crowded marketplace. In parallel they can also leverage the decades of previous development in the other luciferases to jumpstart their engineering of alternative output wavelengths, expression kinetics, and luciferin compatibly. Paired with improvements in bioluminescent detection hardware and modern synesthetic biology engineering tools, it is likely that this renewed age of luciferase engineering will continue to expand the application space for bioluminescent imaging and drive further exploration into the untapped potential of underexplored luciferases.

9. Conclusion

There are a variety of different luciferase systems available for biotechnological applications that can help investigators achieve their experimental goals. The high utility afforded by these enzymes is the result of a rich history of engineering that has enabled them to become versatile research tools. Historically, significant shifts in utility have occurred with the elucidation and introduction of new luciferases, followed by slower, but steady, incremental improvements as they are iteratively engineered to improve their ease of use and expand their functionality. In the context of the historical achievements that have been made with firefly, *Renilla*, *Gaussia*, and *Oplophorus* luciferase, the improvements being made to bacterial luciferase and the recent introduction of fungal luciferase point to promising things to come and give hope that new luciferase systems will continue to be introduced to keep the pace of development strong in the future.

Acknowledgements

Research funding was provided by the U.S. National Institutes of Health under award numbers NIGMS-1R43GM112241, NIGMS-1R41GM116622, NIEHS-2R44ES022567, NIEHS-1R43ES026269, and NIMH-1R43MH118186, and the U.S. National Science Foundation under award number CBET-1530953.

Conflict of interest

S.R., G.S., and D.C. are board members in the for-profit entity 490 BioTech.

Author details

Andrew Kirkpatrick[1], Tingting Xu[2], Steven Ripp[2], Gary Sayler[1,2] and Dan Close[1*]

1 490 BioTech, Inc., Knoxville, Tennessee, USA

2 University of Tennessee, Knoxville, Tennessee, USA

*Address all correspondence to: dan.close@490biotech.com

References

[1] Lee J. Bioluminescence: The first 3000 years. Journal of Siberian Federal University. 2008;**3**(1):194-205

[2] Dubois R. Note sur la physiologie des pyrophores. Comptes Redus des Seances de la Societe de Biologie et de ses Paris. 1885;**8**(2):559-562

[3] McElroy WD. The energy source for bioluminescence in an isolated system. Proceedings of the National Academy of Sciences of the United States of America. 1947;**33**(11):342-345

[4] McElroy WD, Hastings JW, Sonnenfeld V, Coulombre J. The requirement of riboflavin phosphate for bacterial luminescence. Science. 1953;**118**(3066):385-386

[5] Belas R, Mileham A, Cohn D, Hilman M, Simon M, Silverman M. Bacterial bioluminescence: Isolation and expression of the luciferase genes from *Vibrio harveyi*. Science. 1982;**218**(4574):791-793

[6] Kricka LJ, Leach FR. In memoriam Dr. Marlene DeLuca 1987 OM Smith Lecture. Firefly luciferase: Mechanism of action, cloning and expression of the active enzyme. Bioluminescence and Chemiluminescence. 1989;**3**:1-5

[7] Matthews JC, Hori K, Cormier MJ. Purification and properties of *Renilla reniformis* luciferase. Biochemistry. 1977;**16**(1):85-91

[8] Shimomura O, Masugi T, Johnson FH, Haneda Y. Properties and reaction mechanism of the bioluminescence system of the deep-sea shrimp *Oplophorus gracilorostris*. Biochemistry. 1978;**17**(6):994-998

[9] De Wet JR, Wood K, DeLuca M, Helinski DR, Subramani S. Firefly luciferase gene: Structure and expression in mammalian cells. Molecular and Cellular Biology. 1987;7(2):725-737

[10] McNabb DS, Reed R, Marciniak RA. Dual luciferase assay system for rapid assessment of gene expression in *Saccharomyces cerevisiae*. Eukaryotic Cell. 2005;**4**(9):1539-1549

[11] Tannous B, Kim D, Fernandez J, Weissleder R, Breakefield X. Codon-optimized *Gaussia* luciferase cDNA for mammalian gene expression in culture and *in vivo*. Molecular Therapy. 2005;**11**(3):435-443

[12] Kotlobay AA, Sarkisyan KS, Mokrushina YA, Marcet-Houben M, Serebrovskaya EO, Markina NM, et al. Genetically encodable bioluminescent system from fungi. Proceedings of the National Academy of Sciences of the United States of America. 2018;**115**(50):12728-12732

[13] Haddock SH, Moline MA, Case JF. Bioluminescence in the sea. Annual Review of Marine Science. 2010;**2**:443-493

[14] Thorne N, Inglese J, Auld DS. Illuminating insights into firefly luciferase and other bioluminescent reporters used in chemical biology. Chemistry & Biology. 2010;**17**(6):646-657

[15] Widder EA, Falls B. Review of bioluminescence for engineers and scientists in biophotonics. IEEE Journal of Selected Topics in Quantum Electronics. 2014;**20**(2):232-241

[16] Fraga H. Firefly luminescence: A historical perspective and recent developments. Photochemical & Photobiological Sciences. 2008;7(2):146-158

[17] England CG, Ehlerding EB, Cai W. NanoLuc: A small luciferase

is brightening up the field of bioluminescence. Bioconjugate Chemistry. 2016;**27**(5):1175-1187

[18] Close DM, Ripp S, Sayler GS. Reporter proteins in whole-cell optical bioreporter detection systems, biosensor integrations, and biosensing applications. Sensors. 2009;**9**(11):9147-9174

[19] De Wet JR, Wood KV, Helinski DR, DeLuca M. Cloning of firefly luciferase cDNA and the expression of active luciferase in *Escherichia coli*. Proceedings of the National Academy of Sciences of the United States of Amercia. 1985;**82**(23):7870-7873

[20] Marques SM, Esteves da Silva JC. Firefly bioluminescence: A mechanistic approach of luciferase catalyzed reactions. IUBMB Life. 2009;**61**(1):6-17

[21] Viviani VR. The origin, diversity, and structure function relationships of insect luciferases. Cellular and Molecular Life Sciences: CMLS. 2002;**59**(11):1833-1850

[22] Wood KV, Lam YA, McElroy WD. Introduction to beetle luciferases and their applications. Journal of Bioluminescence and Chemiluminescence. 1989;**4**(1):289-301

[23] Koksharov MI, Ugarova NN. Approaches to engineer stability of beetle luciferases. Computational and Structural Biotechnology Journal. 2012;**2**(3):e201204004

[24] Pozzo T, Akter F, Nomura Y, Louie AY, Yokobayashi Y. Firefly luciferase mutant with enhanced activity and thermostability. ACS Omega. 2018;**3**(3):2628-2633

[25] Fujii H, Noda K, Asami Y, Kuroda A, Sakata M, Tokida A. Increase in bioluminescence intensity of firefly luciferase using genetic modification. Analytical Biochemistry. 2007;**366**(2):131-136

[26] Branchini BR, Southworth TL, Khattak NF, Michelini E, Roda A. Red- and green-emitting firefly luciferase mutants for bioluminescent reporter applications. Analytical Biochemistry. 2005;**345**(1):140-148

[27] Branchini BR, Southworth TL, Khattak NF, Murtiashaw MH, Fleet SE, editors. Rational and random mutagenesis of firefly luciferase to identify an efficient emitter of red bioluminescence. In: Genetically Engineered and Optical Probes for Biomedical Applications II. San Jose, California, USA: International Society for Optics and Photonics; 2004

[28] Mofford DM, Reddy GR, Miller SC. Aminoluciferins extend firefly luciferase bioluminescence into the near-infrared and can be preferred substrates over D-luciferin. Journal of the American Chemical Society. 2014;**136**(38):13277-13282

[29] Li J, Chen L, Du L, Li M. Cage the firefly luciferin!—A strategy for developing bioluminescent probes. Chemical Society Reviews. 2013;**42**(2):662-676

[30] Ward WW, Cormier MJ. Energy transfer via protein-protein interaction in Renilla bioluminescence. Photochemistry and Photobiology. 1978;**27**(4):389-396

[31] Lorenz WW, McCann RO, Longiaru M, Cormier MJ. Isolation and expression of a cDNA encoding *Renilla reniformis* luciferase. Proceedings of the National Academy of Sciences of the United States of America. 1991;**88**(10):4438-4442

[32] Srikantha T, Klapach A, Lorenz WW, Tsai LK, Laughlin LA, Gorman JA, et al. The sea pansy *Renilla reniformis* luciferase serves as a

sensitive bioluminescent reporter for differential gene expression in *Candida albicans*. Journal of Bacteriology. 1996;**178**(1):121-129

[33] Mayerhofer R, Langridge WHR, Cormier MJ, Szalay AA. Expression of recombinant *Renilla* luciferase in transgenic plants results in high levels of light emission. The Plant Journal. 1995;7(6):1031-1038

[34] Lorenz WW, Cormier MJ, Okane DJ, Hua D, Escher AA, Szalay AA. Expression of the Renilla reniformis luciferase gene in mammalian cells. Journal of Bioluminescence and Chemiluminescence. 1996;**11**(1):31-37

[35] Bhaumik S, Gamhbir SS. Optical imaging of Renilla luciferase reporter gene exrpession in living mice. Proceedings of the National Academy of Sciences of the United States of Amercia. 2002;**99**(1):377-382

[36] Loening AM, Fenn TD, Wu AM, Gambhir SS. Consensus guided mutagenesis of *Renilla* luciferase yields enhanced stability and light output. Protein Engineering, Design & Selection. 2006;**19**(9):391-400

[37] Liu J, Escher A. Improved assay sensitivity of an engineered secreted *Renilla* luciferase. Gene. 1999;**237**(1):153-159

[38] Loening AM, Wu AM, Gambhir SS. Red-shifted *Renilla reniformis* luciferase variants for imaging in living subjects. Nature Methods. 2007;**4**(8):641-643

[39] Inouye S, Shimomura O. The use of *Renilla* luciferase, *Oplophorus* luciferase, and apoaequorin as bioluminescent reporter protein in the presence of coelenterazine analogues as substrate. Biochemical and Biophysical Research Communications. 1997;**233**(2):349-353

[40] Zhao H, Doyle TC, Wong RJ, Cao Y, Stevenson DK, Piwnica-Worms D, et al. Characterization of coelenterazine analogs for measurements of *Renilla* luciferase activity in live cells and living animals. Molecular Imaging. 2004;**3**(1):43-54

[41] Nishihara R, Suzuki H, Hoshino E, Suganuma S, Sato M, Saitoh T, et al. Bioluminescent coelenterazine derivatives with imidazopyrazinone C-6 extended substitution. Chemical Communications. 2015;**51**(2):391-394

[42] Nishihara R, Abe M, Nishiyama S, Citterio D, Suzuki K, Kim SB. Luciferase-specific coelenterazine analogues for optical contamination-free bioassays. Scientific Reports. 2017;**7**(1):908

[43] Paulmurugan R, Gambhir SS. Monitoring protein-protein interactions using split synthetic *Renilla* luciferase protein-fragment-assisted complementation. Analytical Chemistry. 2003;**75**(7):1584-1589

[44] Jiang Y, Bernard D, Yu Y, Xie Y, Zhang T, Li Y, et al. Split *Renilla* luciferase protein fragment-assisted complementation (SRL-PFAC) to characterize Hsp90-Cdc37 complex and identify critical residues in protein/protein interactions. The Journal of Biological Chemistry. 2010;**285**(27):21023-21036

[45] Wang J, Guo W, Long C, Zhou H, Wang H, Sun X. The split Renilla luciferase complementation assay is useful for identifying the interaction of Epstein-Barr virus protein kinase BGLF4 and a heat shock protein Hsp90. Acta Virologica. 2016;**60**(1):62-70

[46] Kim SB, Ozawa T, Watanabe S, Umezawa Y. High-throughput sensing and noninvasive imaging of protein nuclear transport by using reconstitution of split *Renilla* luciferase. Proceedings of the National Academy

of Sciences of the United States of America. 2004;**101**(32):11542-11547

[47] Kaihara A, Kawai Y, Sato M, Ozawa T, Umezawa Y. Locating a protein–protein interaction in living cells via split *Renilla* luciferase complementation. Analytical Chemistry. 2003;**75**(16):4176-4181

[48] Stefan E, Aquin S, Berger N, Landry CR, Nyfeler B, Bouvier M, et al. Quantification of dynamic protein complexes using *Renilla* luciferase fragment complementation applied to protein kinase a activities *in vivo*. Proceedings of the National Academy of Sciences. 2007;**104**(43):16916-16921

[49] Lund CH, Bromley JR, Stenbaek A, Rasmussen RE, Scheller HV, Sakuragi Y. A reversible *Renilla* luciferase protein complementation assay for rapid identification of protein-protein interactions reveals the existence of an interaction network involved in xyloglucan biosynthesis in the plant Golgi apparatus. Journal of Experimental Botany. 2015;**66**(1):85-97

[50] Hatzios SK, Ringgaard S, Davis BM, Waldor MK. Studies of dynamic protein-protein interactions in bacteria using *Renilla* luciferase complementation are undermined by nonspecific enzyme inhibition. PLoS One. 2012;**7**(8):e43175

[51] Tannous BA. *Gaussia* luciferase reporter assay for monitoring biological processes in culture and *in vivo*. Nature Protocols. 2009;**4**(4):582-591

[52] Chung E, Yamashita H, Au P, Tannous BA, Fukumura D, Jain RK. Secreted Gaussia luciferase as a biomarker for monitoring tumor progression and treatment response of systemic metastases. PLoS One. 2009;**4**(12):e8316

[53] Badr CE, Hewett JW, Breakefield XO, Tannous BA. A highly sensitive assay for monitoring the secretory pathway and ER stress. PLoS One. 2007;**2**(6):e571

[54] Suzuki T, Usuda S, Ichinose H, Inouye S. Real-time bioluminescence imaging of a protein secretory pathway in living mammalian cells using *Gaussia* luciferase. FEBS Letters. 2007;**581**(24):4551-4556

[55] Bruce BJ, Szent-Gyorgyi CS. Luciferases, fluorescent proteins, nucleic acids encoding the luciferases and fluorescent proteins and the use thereof in diagnostics, high throughput screening and novelty items. U.S. Patent 6,232,107. 2001

[56] Shao N, Bock R. A codon-optimized luciferase from *Gaussia princeps* facilitates the in vivo monitoring of gene expression in the model alga *Chlamydomonas reinhardtii*. Current Genetics. 2008;**53**(6):381-388

[57] Enjalbert B, Rachini A, Vediyappan G, Pietrella D, Spaccapelo R, Vecchiarelli A, et al. A multifunctional, synthetic *Gaussia princeps* luciferase reporter for live imaging of *Candida albicans* infections. Infection and Immunity. 2009;**77**(11):4847-4858

[58] Andreu N, Zelmer A, Fletcher T, Elkington PT, Ward TH, Ripoll J, et al. Optimisation of bioluminescent reporters for use with mycobacteria. PLoS One. 2010;**5**(5):e10777

[59] Wille T, Blank K, Schmidt C, Vogt V, Gerlach RG. *Gaussia princeps* luciferase as a reporter for transcriptional activity, protein secretion, and protein-protein interactions in *Salmonella enterica* serovar typhimurium. Applied and Environmental Microbiology. 2012;**78**(1):250-257

[60] Kim SB, Sato M, Tao H. Split *Gaussia* luciferase-based bioluminescence template for tracing protein dynamics in living cells. Analytical Chemistry. 2009;**81**(1):67-74

[61] Degeling MH, Bovenberg MSS, Lewandrowski GK, de Gooijer MC, Vleggeert-Lankamp CLA, Tannous M, et al. Directed molecular evolution reveals *Gaussia* luciferase variants with enhanced light output stability. Analytical Chemistry. 2013;**85**(5):3006-3012

[62] Welsh JP, Patel KG, Manthiram K, Swartz JR. Multiply mutated *Gaussia* luciferases provide prolonged and intense bioluminescence. Biochemical and Biophysical Research Communications. 2009;**389**(4):563-568

[63] Maguire CA, Deliolanis NC, Pike L, Niers JM, Tjon-Kon-Fat LA, Sena-Esteves M, et al. Gaussia luciferase variant for high-throughput functional screening applications. Analytical Chemistry. 2009;**81**(16):7102-7106

[64] Remy I, Michnick SW. A highly sensitive protein-protein interaction assay based on *Gaussia* luciferase. Nature Methods. 2006;**3**(12):977-979

[65] Luker KE, Luker GD. Split *Gaussia* luciferase for imaging ligand-receptor binding. Methods in molecular biology (Clifton, NJ). 2014;**1098**:59-69

[66] Yamaguchi I. *Oplophorus* oxyluciferin and a model luciferin compound biologically active with *Oplophorus luciferase*. The Biochemical Journal. 1975;**151**(1):9-15

[67] Inouye S, Watanabe K, Nakamura H, Shimomura O. Secretional luciferase of the luminous shrimp *Oplophorus gracilirostris*: cDNA cloning of a novel imidazopyrazinone luciferase. FEBS Letters. 2000;**481**(1):19-25

[68] Hall MP, Unch J, Binkowski BF, Valley MP, Butler BL, Wood MG, et al. Engineered luciferase reporter from a deep sea shrimp utilizing a novel imidazopyrazinone substrate. ACS Chemical Biology. 2012;7(11): 1848-1857

[69] Chen Y, Wang L, Cheng X, Ge X, Wang P. An ultrasensitive system for measuring the USPs and OTULIN activity using Nanoluc as a reporter. Biochemical and Biophysical Research Communications. 2014;**455**(3-4):178-183

[70] Lackner DH, Carré A, Guzzardo PM, Banning C, Mangena R, Henley T, et al. A generic strategy for CRISPR-Cas9-mediated gene tagging. Nature Communications. 2015;**6**:10237

[71] Zhao J, Nelson TJ, Vu Q, Truong T, Stains CI. Self-assembling NanoLuc luciferase fragments as probes for protein aggregation in living cells. ACS Chemical Biology. 2015;**11**(1):132-138

[72] Meighen EA. Molecular biology of bacterial bioluminescence. Microbiological Reviews. 1991;**55**:123-142

[73] Engebrecht J, Nealson K, Silverman M. Bacterial bioluminescence: Isolation and genetic analysis of functions from *Vibrio fischeri*. Cell. 1983;**32**:773-781

[74] Fisher AJ, Raushel FM, Baldwin TO, Rayment I. Three-dimensional structure of bacterial luciferase from *Vibrio harveyi* at 2.4. ANG. Resolution. Biochemistry. 1995;**34**(20):6581-6586

[75] Engebrecht J, Simon M, Silverman M. Measuring gene expression with light. Science. 1985;**227**(4692):1345-1347

[76] King JMH, Digrazia PM, Applegate B, Burlage R, Sanseverino J, Dunbar P, et al. Rapid, sensitive bioluminescent reporter technology for naphthalene exposure and biodegradation. Science. 1990;**249**(4970):778-781

[77] Shaw JJ, Kado CI. Development of a *Vibrio* bioluminescence gene set to monitor phytopathogenic bacteria during the ongoing disease process in a nondisruptive manner. Nature Biotechnology. 1986;**4**(6):560-564

[78] de Weger L, Dunbar P, Mahafee F, Lugtenberg B, Sayler G. Use of bioluminescence markers to detect *Pseudomonas* spp. in the rhizosphere. Applied and Environmental Microbiology. 1991;57:3641-3644

[79] Ripp S, Nivens DE, Ahn Y, Werner C, Jarrell J, Easter JP, et al. Controlled field release of a bioluminescent genetically engineered microorganism for bioremediation process monitoring and control. Environmental Science & Technology. 2000;**34**(5):846-853

[80] Kuklin NA, Pancari GD, Tobery TW, Cope L, Jackson J, Gill C, et al. Real-time monitoring of bacterial infection *in vivo*: Development of bioluminescent staphylococcal foreign-body and deep-thigh-wound mouse infection models. Antimicrobial Agents and Chemotherapy. 2003;**47**(9):2740-2748

[81] Almashanu S, Musafia B, Hadar R, Suissa M, Kuhn J. Fusion of *luxA* and *luxB* and its expression in *Escherichia coli*, *Saccharomyces cerevisiae* and *Drosophila melanogaster*. Journal of Bioluminescence and Chemiluminescence. 1990;**5**(1):89-97

[82] Kirchner G, Roberts JL, Gustafson GD, Ingolia TD. Active bacterial luciferase from a fused gene: Expression of a *Vibrio harveyi luxAB* translational fusion in bacteria, yeast and plant cells. Gene. 1989;**81**(2):349-354

[83] Olsson O, Koncz C, Szalay AA. The use of the *luxA* gene of the bacterial luciferase operon as a reporter gene. Molecular & General Genetics. 1988;**215**(1):1-9

[84] Gupta RK, Patterson SS, Ripp S, Sayler GS. Expression of the *Photorhabdus luminescens lux* genes (*luxA*, *B*, *C*, *D*, and *E*) in *Saccharomyces cerevisiae*. FEMS Yeast Research. 2003;**4**(3):305-313

[85] Patterson SS, Dionisi HM, Gupta RK, Sayler GS. Codon optimization of bacterial luciferase (*lux*) for expression in mammalian cells. Journal of Industrial Microbiology & Biotechnology. 2005;**32**(3):115-123

[86] Close DM, Hahn R, Patterson SS, Ripp S, Sayler GS. Comparison of human optimized bacterial luciferase, firefly luciferase, and green fluorescent protein for continuous imaging of cell culture and animal models. Journal of Biomedical Optics. 2011;**16**(4):e12441

[87] Xu T, Ripp S, Sayler G, Close D. Expression of a humanized viral 2A-mediated *lux* operon efficiently generates autonomous bioluminescence in human cells. PLoS One. 2014;**9**(5):e96347

[88] Gregor C, Gwosch KC, Sahl SJ, Hell SW. Strongly enhanced bacterial bioluminescence with the *ilux* operon for single-cell imaging. Proceedings of the National Academy of Sciences. 2018;**115**(5):962-967

[89] Welham PA, Stekel DJ. Mathematical model of the *Lux* luminescence system in the terrestrial bacterium *Photorhabdus luminescens*. Molecular BioSystems. 2009;**5**(1):68-76

[90] Yagur Kroll S, Belkin S. Upgrading bioluminescent bacterial bioreporter performance by splitting the *lux* operon. Analytical and Bioanalytical Chemistry. 2011;**400**(4):1071-1082

[91] Airth R, McElroy W. Light emission from extracts of luminous fungi. Journal of Bacteriology. 1959;**77**(2):249-250

[92] Oliveira AG, Stevani CV. The enzymatic nature of fungal bioluminescence. Photochemical & Photobiological Sciences. 2009;**8**(10):1416-1421

[93] Oliveira AG, Desjardin DE, Perry BA, Stevani CV. Evidence that a single

bioluminescent system is shared by all known bioluminescent fungal lineages. Photochemical & Photobiological Sciences. 2012;**11**(5):848-852

[94] Oba Y, Stevani CV, Oliveira AG, Tsarkova AS, Chepurnykh TV, Yampolsky IV. Selected least studied but not forgotten bioluminescent systems. Photochemistry and Photobiology. 2017;**93**(2):405-415

Chapter 2

Protein-Protein Interaction Assays Using Split-NanoLuc

Yuki Ohmuro-Matsuyama and Hiroshi Ueda

Abstract

Protein-protein interaction assays are fundamental to basic biology, drug discovery, diagnostics, screening, and immunoassays. Protein-fragment complementation (PCA) is one of such useful protein-protein interaction assays. PCA when performed using luciferase is a reversible approach, whereas when performed using green fluorescent protein analogs is an irreversible approach. The NanoLuc technology developed in 2012 utilizes a small and structurally robust luciferase that is capable of producing very bright luminescence. NanoLuc PCA has been used to detect many protein-protein interactions and for screening purposes. Methods developed from NanoLuc PCA include the HiBiT technology and NanoLuc ternary technology. These novel technologies are promising in various fields and further developments are anticipated.

Keywords: NanoLuc, PCA, NanoBiT, protein-protein interaction, HiBiT, NanoLuc ternary technology

1. Introduction

It is predicted that there are 150,000–650,000 protein-protein interactions in the human interactome [1–3]. Protein-protein interaction assays have been developed and used for studies on basic biology, drug discoveries, diagnostics, screenings, and immunoassays.

In 1994, the first protein-fragment complementation assay (PCA) was developed using split ubiquitin [4]. PCA typically uses two-split reporter proteins that are fused to the target proteins. The interaction leads to the association of the fragments and the subsequent reconstitution of the full-length structure from the two fragments (**Figure 1**) [5–8]. More recently, fluorescent proteins and luciferase enzymes have been widely utilized for innovative PCAs. Reversible PCAs generally utilize enzymes, and the exceptions are two fluorescent proteins IFP1.4 and UnaG [9, 10]. Most other PCA systems that use fluorescent proteins, including green fluorescent protein (GFP) analogs, show irreversible behavior. In such irreversible assays, once the full-length structure is reconstituted, it is difficult to separate them into the two fragments when dissociation occurs after the interaction. On the contrary, in the reversible PCA systems, both interaction and dissociation can be detected. Therefore, PCA systems using enzymes, such as luciferase, are more suitable to detect the spatiotemporal dynamics of protein-protein interactions. However, until recently, the luminescent signal is significantly weaker than the fluorescent signal.

Recently, a novel luciferase enzyme, NanoLuc, and its furimazine substrate were developed [11, 12]. NanoLuc is small (19 kDa) and structurally stable, and produces

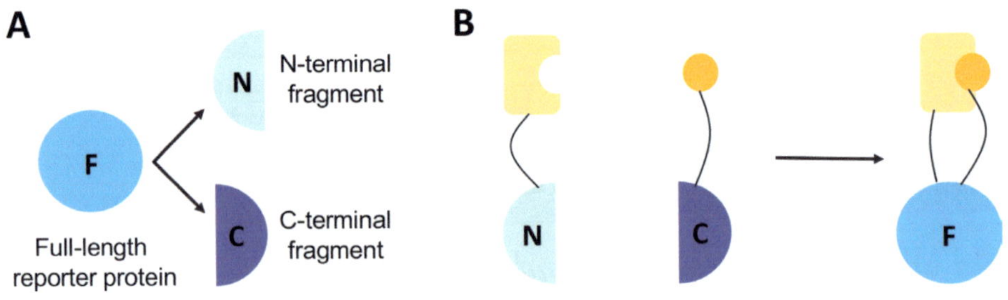

Figure 1.
Basic principle of PCA. (A) The reporter protein (F) is separated into two fragments (N and C). (B) N and C are fused to target protein (left). When the interaction occurs, N and C move to the neighboring position, and the full-length reporter protein is reconstituted.

very bright luminescence. Based on this attractive enzyme, PCA systems were developed [13, 14]. This innovation on the NanoLuc PCA improves the luminescent signal, which is markedly better than the conventional PCA signal obtained using other luciferases. Herein, we will focus on the new PCA technology and its application, and further discuss potential improvements in the system.

2. PCA using NanoLuc

Verhoef et al. constructed a PCA system using NanoLuc [14]. They made several pairs of NanoLuc fragments by cutting at several loop regions, and selected a pair comprised of the N-terminal 52-amino acid (aa) fragment and the C-terminal 119-aa fragment (**Figure 2A**). These fragments were used to successfully detect the interaction between the transactivation domain fragment of p53 and Mdm2.

At almost the same time, Dixon et al. developed another NanoLuc-based PCA system designated *Nano*Luc *Bi*nary *T*echnology (NanoBiT) [13]. This was devised by first identifying a dissection site from 90 candidate sites. An 18-kDa N-terminal fragment and 13-aa C-terminal fragment were selected. The K_D value between these fragments was 6 μM. This low affinity was suitable for PCA, but their use was hampered by the very low stability of the N-terminal fragment. The sequence of the N-terminal fragment was optimized from an N-terminal library containing 15,000 variants. The optimization increased the luminescent signal by 300-fold when the two fragments were interacting, which was 37% that of the wild-type NanoLuc. However, the affinity between the N- and C-terminal fragments became too strong for PCA (K_D = 900 nM). As a next step, the sequence of the C-terminal peptide was optimized from 350 variants. Finally, two fragments were obtained. They were designated LgBiT (18 kDa) and SmBiT (11 aa). These exhibited significantly low

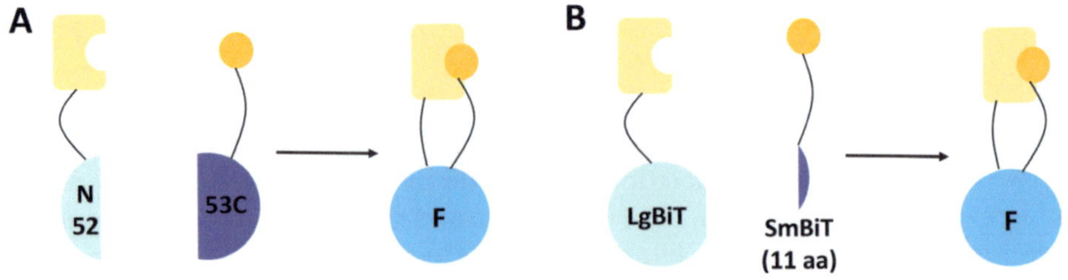

Figure 2.
Two systems of NanoLuc PCA. (A) Verhoef et al. separated NanoLuc into N-terminal 52-aa fragment and C-terminal 119-aa fragment. (B) Dixon et al. separated NanoLuc into the large fragment, LgBiT (18 kDa), and the small fragment, SmBiT (11 aa).

affinity (K_D > 10 μM) and high luminescent intensity. The very bright signal and remarkably high signal/background ratio obtained enabled the quantitative detection of several interactions. Furthermore, the luminescent signals were capable of rapid change and were reversible depending on changing interactions (**Figure 2B**).

3. Application of NanoBiT for analysis of protein-protein interaction

In spite of recent appearance, NanoBiT has been already used to analyze several protein-protein interactions. Elevation of plasma triglycerides causes various metabolic diseases. These triglycerides are digested by lipoprotein lipase [15–19]. Chi et al. used NanoBiT to demonstrate the association between lipoprotein lipase and angiopoietin-like 3 (ANGPTL3) induced by ANGPTL8 [20]. They further described that the association inhibits the digestion activity of lipoprotein lipase.

Guanine nucleotide-binding (G) protein-coupled receptors (GPCRs) bind G proteins or β-arrestins, and initiate several cellular signaling events. Regulator of G protein signaling (RGS) proteins regulate G proteins. The regulation has been implicated in several disease states, including various cancers, Parkinson's disease, and cardiomyopathy [21–25]. Several reports have described the use of NanoBiT to analyze the mechanisms of GPCRs. These included the interaction of several sets of RGS proteins and G proteins [26] and the interaction between the galanin receptor 2 GPCR and β-arrestin2 [27]. Furthermore, the LgBiT-fused galanin receptor 2 was modified with a fluorescent dye, and the conformational changes of galanin receptor induced by the binding of ligands, including galanin, spexin, and Fmoc-dA4-dQ14, were analyzed by bioluminescence resonance energy transfer (BRET). Stome et al. applied NanoBiT to analyze the interaction between the GPCR adenosine receptor 3 and β-arrestin2, and observed that the 1-deoxy-1-[6-[[(3-iodophenyl)methyl] amino]-9H-purin-9-yl]-N-methyl-β-D-ribofuranuronamide (2-CI-IB-MECA) agonist recruited β-arrestin2 [28]. In addition, the authors described the importance of the phosphorylation site of adenosine receptor 3 for the association. The site was implicated as a potential clinical target. Melanocortin receptors are also categorized as GPCRs. Melanocortin 4 receptor (MC4R) binds one of melanocortins α-melanocyte-stimulating hormone (α-MSH), which is considered is important in obesity. Habara et al. isolated melanocortin receptor 4 and its regulator Melanocortin 2 receptor accessory protein 2 (MRAP2) from cats and analyzed the heterodimerization of these proteins [29]. Leory et al. characterized several mutants of Janus Kinase 2 in signaling by a transmembrane cytokine receptor, erythropoietin receptor [30].

Interactions with other membrane proteins implicated as important drug targets were also analyzed by NanoBiT. Folding and steric hindrance are problematic for many membrane proteins. The small size of SmBiT could eliminate these problems. O'Neil et al. revealed the amino acids of NADPH that were important for the interaction with p22 using NanoBiT [31]. Chaudhri et al. applied NanoBiT to analyze the association between programmed death ligand 1 (PD-L1) and B7–1 [32]. Peptide hormone, a member of the relaxin family of peptides, participates in reproduction, food intake, stress response, and glucose homeostasis [33–36]. Hu et al. demonstrated the association between the relaxin family peptide receptors peptide 3 and peptide 4 using NanoBiT [37]. The same group further reported that the interaction is electrostatic by analyzing the association between several mutant ligands and their receptors [38]. Equilibrative nucleoside transporters regulate the levels of adenosine and hypoxanthine level, and are crucial in purinergic signaling in the central nervous system, cardiovascular and renal systems, and in pathophysiological conditions including myocardial ischemia, inflammation, and diabetic nephropathy [39–41]. Grañe-Boladeras et al. analyzed the homo- and hetero-oligomerization of ENT1 and ENT2, and revealed that the phosphorylation by protein kinase C promotes oligomerization [42].

4. Application of NanoBiT for screening

Several groups have successfully used NanoBiT in highly accurate drug screening, including illegal drugs [43, 44]. In the latter studies, β-arrestin2 was fused to SmBiT, and the CB1 and CB2 GPCRs of cannabinoid (the neurologically active component of cannabis) were fused to LgBiT. As cannabinoid induces the interaction between β-arrestin2 and these receptors, the luminescent intensity was increased by adding synthetic cannabinoids and their metabolites. The synthetic cannabinoids and metabolites were detected in subnanomolar concentrations in authentic urine samples with an accuracy rate of 73%.

Next, the authors tried to detect synthetic opioids, which act similarly to heroin or morphine. The μ-opioid receptor and β-arrestin2, which interact in the presence of opioid, were fused to LgBiT and SmBiT, respectively [45]. The system was nearly 100% successful in detecting subnanomolar levels of the synthetic opioids in blood samples.

Aggregation of TDP (transactivating response region DNA binding protein)-43 occurs in approximately 95% of amyotrophic lateral sclerosis patients [46, 47]. Oberstadt et al. constructed a screening system for inhibitors of aggregation by the fusion between the LgBiT and SmBiT probes and TDP-43 [48]. Aurorafin, chelerythrine, and riluzole were identified as inhibitors from the Library of Pharmacologically Active Compounds (LOPAC1280).

Stomes et al. selected agonists of the interaction between adenosine receptor 3 and β-arrestin2 and revealed the structural features of the selected ligands [28].

The NanoBiT screening system is not only effective for drug screening but can be valuable to screen enzyme substrates. Peptide ligases, which can connect two polypeptides, are powerful tools for protein engineering [49–52]. Li et al. performed the screening of substrates of the peptide ligase Sortase A by fusing this enzyme to SmBiT and the candidate peptides to LgBiT [53]. In addition to known substrate sequences, they rapidly identified some previously unknown substrates with varying activities. In addition, the measurement was very stable, and the signal was maintained for more than 16 h.

5. Application of NanoBiT using self-assembling NanoLuc fragments

Self-assembling NanoLuc fragments have been used to detect protein aggregation, to detect the edited protein by CRISPR/Cas9, to monitor viral entry, release, and propagation, and to analyze clathrin-dependent internalization (**Figure 3**).

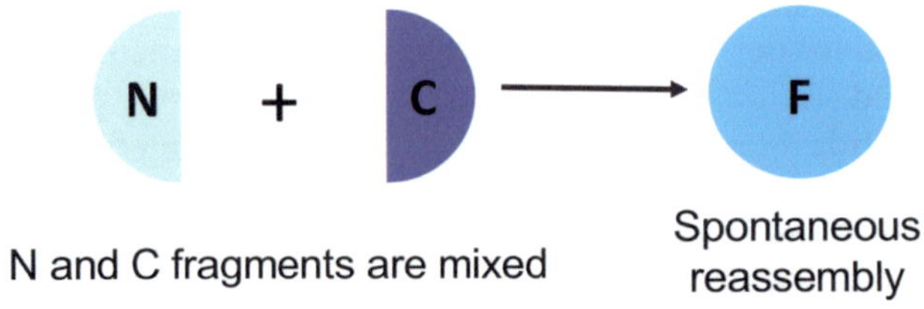

Figure 3.
Scheme for protein fragments self-assembly. Since the affinity between N and C is high, the full-length reporter protein is reconstituted by just mixing N and C.

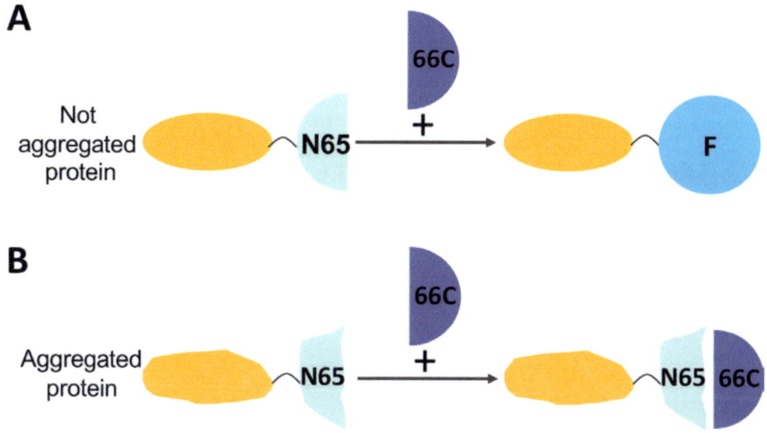

Figure 4.
Self-assembling Nluc fragment as a probe for protein aggregation. (A) N65 is fused to the target protein. When the fusion protein is not aggregated, NanoLuc is reconstituted by the addition of 66C. (B) When the target protein is aggregated, N65 is also aggregated, and, then, the reassembly does not occur.

The first description of the use of self-assembling NanoLuc fragments was provided by Zhao et al [54]. NanoLuc was separated into two fragments, N65 (1–65 aa) and 66C (66–171 aa). NanoLuc was rapidly reconstituted when N65 and 66C were mixed. Next, N65 was fused to the target proteins. When the target protein was soluble, N65, which had the correct structure, could reassemble with 66C, resulting in recovery of the luminescence. On the other hand, the insoluble target protein did not induce the recovery of the luminescence, because the aggregated N65 could not assemble with 66C (**Figure 4**). The aggregations of amyloid-β mutants were assessed using the system. Similar monitoring systems of protein aggregation using split-GFP and conventional split-luciferase systems had been previously reported [55–60]. However, a time lag occurred for the chromophore formation in the split-GFP system, and other luciferases were relatively unstable compared with NanoLuc. Zhao et al. succeeded the robust measurement of amyloid-β in this study.

Other self-assembling NanoLuc fragments were described [13]. The SmBiT sequence was optimized using peptides with different affinities to LgBiT. Of the candidates, the HiBiT peptide displayed high affinity (K_D = 700 pM) although the affinity of SmBiT was very low (K_D > 100 μM). HiBiT (11 aa) and LgBiT assembled spontaneously, allowing the construction of NanoLuc. HiBiT is a useful tag due to the small size as further described below.

In one of the split-GFP systems, GFP was split into two fragments [57]. The C-terminal fragment of GFP contains 16 aa (GFP11). Waldo et al. found that this and the other fragment (GFP1–10) expressed in the cell assembled spontaneously, and the GFP fluorescence was recovered. Leonetti et al. described the synthesis of the donor DNA templates encoding GFP11 and the tagging of endogenous proteins using CRISPR/Cas9 (**Figure 5A**) [61]. The formation of full-length GFP was induced by coexpression with GFP1–10, and the tagging endogenous protein by GFP11 could be detected. Instead of GFP11, Schwinn et al. used HiBiT as the tag for endogenous proteins and were successful in achieving the highly efficient integration and monitoring of the expression dynamics of the tagging proteins without the time lag, which occurs in the split-GFP system due to the chromophore formation (**Figure 5B**) [62].

Ryes-Alaraz et al. analyzed the internalization of galanin receptor 2, which is dependent on the binding of the endogenous ligand, using an HiBiT-fused galanin receptor 2 [27]. LgBiT could bind to the HiBiT-fused receptor on the cell surface.

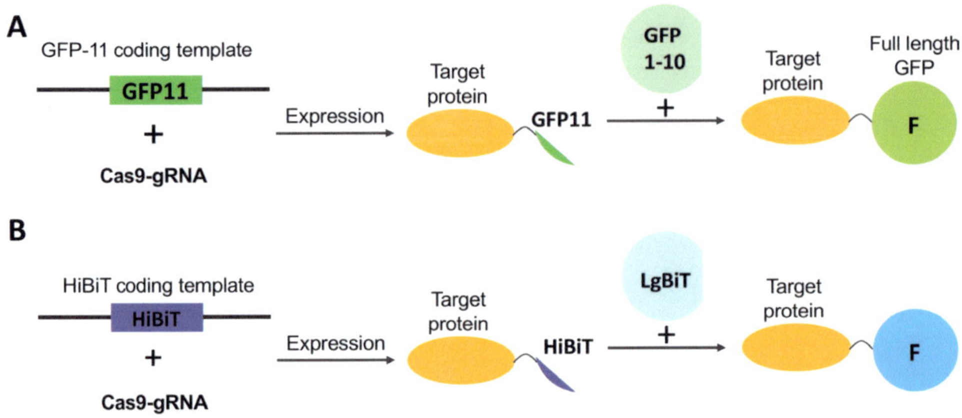

Figure 5.
Application of self-assembling fragment for Crispr-Cas9 system. (A) GFP11-encoding template is inserted at the end of the genome encoding target protein by Crispr-Cas9 system. GFP11-fused target protein is expressed. By the addition of GFP1–10, full-length GFP is reconstituted and the target protein was detected. (B) HiBiT is used instead of GFP11, and higher sensitivity is attained.

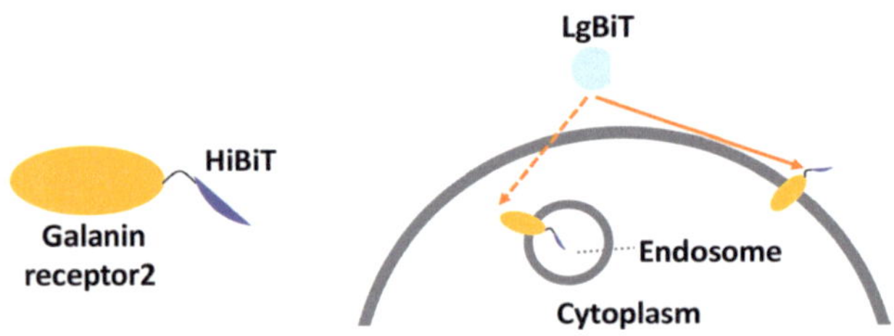

Figure 6.
Application of self-assembling fragment for the analysis of internalization. LgBiT can bind to HiBiT-galanin receptor 2 on the cell membrane, while it cannot access HiBiT-galanin receptor 2 in the endosome.

However, LgBiT could not bind to the HiBiT-fused receptor in cells due to the impermeability of LgBiT to cells (**Figure 6**).

This technology has often been used to quantify targets. Oh-hashi et al. used HiBiT to quantify the expression of transcription factor ATF4 that was induced by endoplasmic reticulum stress [63]. Sasaki et al. developed a quantitative detection system of viral entry and release using HiBiT fused to subviral particles and flavivirus-like particles of West Nile virus [64]. Tamura et al. constructed recombinant viruses carrying HiBiT [65]. Viral amplification and propagation were rapid and comparable with the parental viruses, due to the small size of HiBiT. The techniques proved useful to study the viral life cycle and pathogenesis.

6. NanoLuc ternary technology

In PCA, the reporter protein is generally separated into two fragments. To our knowledge, PCA using 3-split reporter protein was first reported by Cabantous et al. [66]. In the study, GFP was split into two peptides, GFP10 and GFP11, and the remaining part. The two peptides were each fused to an interacting partner. When the interaction occurred, the peptides came into close proximity with one another and then assembled to form the full length of GFP with the remaining part (**Figure 7**).

Possibly inspired by this GFP ternary technology, Dixon et al. developed the NanoLuc ternary technology, NanoLuc is consisted of 11 β-strands [67]. The

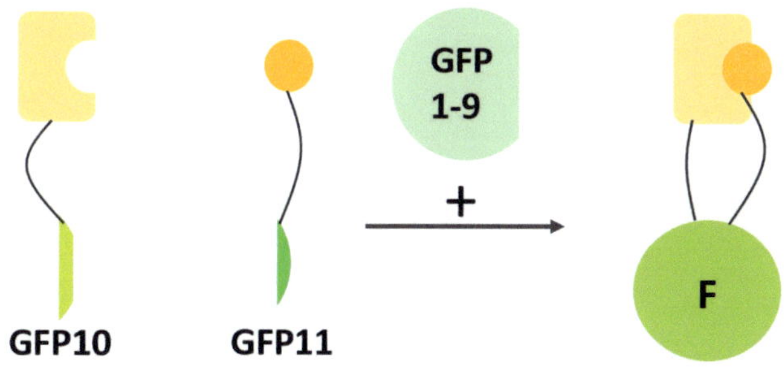

Figure 7.
GFP ternary split technology for the detection of protein-protein interaction. The two small fragments GFP10 and GFP11 are fused to the interacting proteins, respectively. When the interaction occurs, GFP is reconstituted from GFP10, GFP11, and externally added GFP1–9.

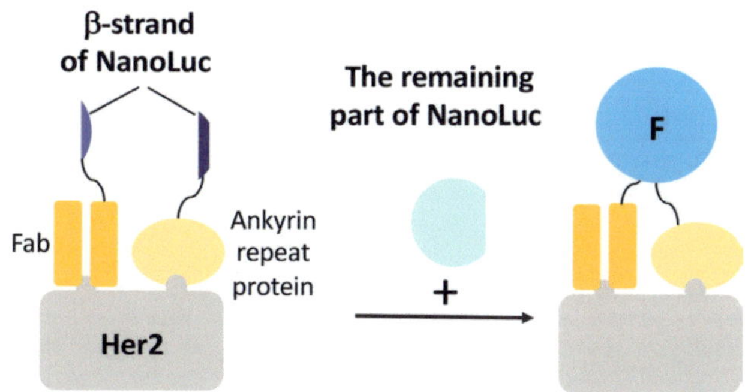

Figure 8.
Sandwich immunoassay based on NanoLuc ternary technology developed by Dixon et al. The two β-strands are fused to a Fab and an ankyrin repeat protein, respectively, which bind to two distant parts of Her2 protein. NanoLuc was reconstituted from the two strands and externally added remainder of NanoLuc.

authors dissected two β-strands and the remaining part. Each strand was fused to a Fab fragment of an antibody and an ankyrin repeat protein, which bound to distinct areas of the cancer marker, HER2. When both antibodies recognized HER2, the two strands came close together and the full length of NanoLuc was reconstituted from the three fragments of NanoLuc (**Figure 8**). The sensitivity was similar to the sensitivity detected using the commercially available AlphaLISA HER2 kit (Perkin Elmer) and NanoBiT. Furthermore, the detectable concentration range of HER2 was broader compared to the range detected by NanoBiT.

At almost the same time, we developed the NanoLuc ternary technology for use as an open-sandwich immunoassay (OS-IA), because OS-IA could not be performed using NanoBiT [68]. For OS-IA, two antigen-binding regions, the heavy-chain variable region (V_H) and the light-chain variable region (V_L), were isolated from the full-length antibody. OS-IA is based on the antigen-dependent interaction affinity between V_H and V_L, which is dependent on the antigen (**Figure 9**) [69]. The advantage of OS-IA is that small antigens can be noncompetitively detected with high sensitivity. V_H and V_L were fused to LgBiT and SmBiT. However, the signal was not increased by the addition of the small peptide antigen (7 aa) named BGP-C7. We suspected that fusion with LgBiT sterically hindered the interaction, or prevented the folding of these antibody fragments due to the relatively large size of LgBiT.

The next step was to split LgBiT in two. The C-terminal strand (11 aa) was named LcBiT, and the remaining part was named LnBiT. LcBiT and SmBiT were fused to V_H and V_L, respectively. When LnBiT, V_H-LcBiT, and VL-SmBiT were mixed, the signal was increased depending on the concentration of BGP-C7. The

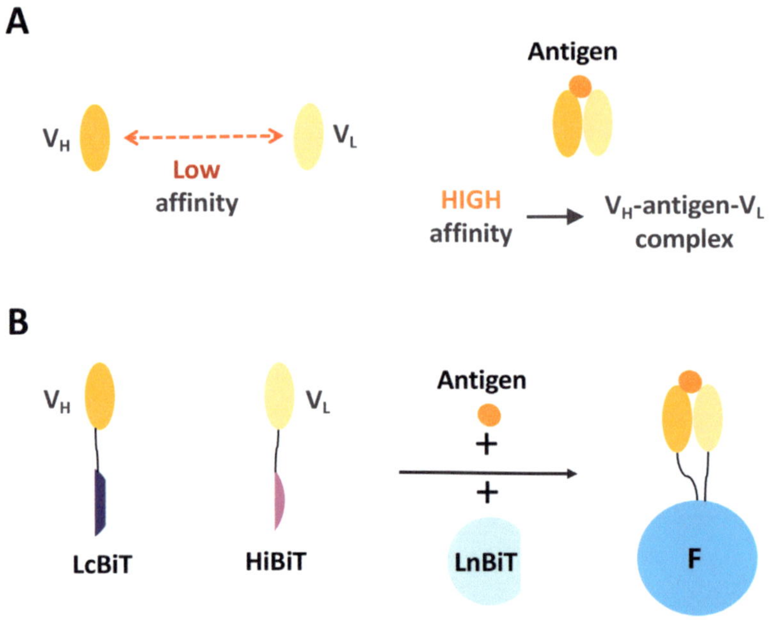

Figure 9.
Open sandwich immunoassay (OS-IA) and NanoLuc ternary technology. (A) Principle of OS-IA. Small antigens (MW < 1000) can be noncompetitively detected. (B) Noncompetitive detection of small antigen by NanoLuc ternary technology. Visual detection was possible when sufficient amount of antigen was present.

background signal without BGP-C7 was lower than the background signal of V_H-LgBiT and V_L-SmBiT.

Next, the signal was enhanced by optimizing the sequence of SmBiT. The signal was increased 288-fold using the sequence, which has higher affinity to LgBiT. The enhancement was high enough to permit detection by the naked eye. The detection limit of BGP-C7 was comparable with the limit detected by OS-ELISA. Furthermore, the strong signal was maintained for more than 1 h.

The small tags of the NanoLuc ternary system have proven to be very useful when both target proteins have complex structures. Furthermore, the system exhibits a robust and bright signal.

7. Discussion and conclusion

The described NanoLuc binary and ternary technologies are superior compared with other PCAs using other luciferase enzymes. The signals obtained were almost the strongest among the signals of PCAs using luciferase enzymes. The most important advantage is the small size of the fusion tags, SmBiT and LcBiT.

As mentioned in Section 4, NanoBiT has been used as a screening tool. In several studies, SmBiT was fused to membrane proteins, which are important targets of drug discovery. The brightness of NanoLuc increased the hit ratio, and SmBiT was validated as a tag for the fusion with proteins having complex structures, such as membrane proteins. Although the measurements were very accurate, researchers should pay attention to the influence of low-molecular weight compounds on enzymatic activity [70–72]. Some compounds increase enzymatic activity, while others decrease it. The TurboLuc system reported by Audi et al. is somewhat smaller in molecular weight (16 kDa) than NanoLuc [73]. Compared with NanoLuc and Firefly luciferase, the activity of TurboLuc was less affected by low-molecular weight compounds. Ho et al. examined the influence of 42,460 PubChem compounds on enzymatic activities of several luciferases [74]. NanoLuc, *Renilla* luciferase, firefly luciferase, and *Gaussia* luciferase were affected by 2.7, 10, 4, and 0.02% of the compounds, respectively. The relationship between the chemical similarity and the inhibition profile showed that the compounds varied depending on the luciferase used. While NanoLuc has several advantages for screening, in some cases, researchers should select other enzymes or more than two enzymes. Furthermore, we previously cautioned using a mathematical model that the comparison between the affinity of interacting proteins and the signal detected by luciferase-based PCA can cause misinterpretation of the quantitation [75]. In addition, we suggested that the geometry of the interacting proteins influences the luminescent signal. In other words, the structures of the interacting proteins can affect the reconstitution of luciferase. Quantitative measurement with PCA using luciferase is possible, but careful examinations are needed.

One of the other problems is the unstable luminescence in cells. When we use the standard furimazine ester for live cell assay, the light intensity decreases within 1 h, and it will not be suitable for large screenings in cellulo. Recently, live cell substrates with longer half-life (Vivazine and Endurazine, Promega) have become available. These substrates can maintain the luminescent signal for several hours, although the peak luminescent intensity is significantly lower compared with that detected using the conventional furimazine ester. However, the luminescent intensity of unmodified furimazine can be maintained for several hours in vitro. Oxygen in cells and the culture medium will also be an important factor for the stable luminescence because the NanoLuc-catalyzed reaction requires oxygen, especially when the light emitted is strong. The last problem is the prices of these substrates, which tend to be costly especially when larger scale screening is intended.

Detection of the interaction among more than three proteins, and the simultaneous detection of more than two interactions will become more important in future. For the detection among three proteins, the combination of NanoBiT and BRET might be useful. The brightness of NanoLuc often disturbs simultaneous detection using both NanoLuc and other luciferases. Therefore, NanoLuc inhibitors were developed [76]. After the measurement of NanoLuc luminescence, the luminescence can be diminished by the inhibitors, which enables detection of the luminescence of another luciferase. For the simultaneous detection by multicolors, several color variants of eNano-Lantern (a fusion protein of NanoLuc and fluorescent protein) can be applied to NanoBiT. In eNano-Lantern, the luminescence at longer wavelength can be observed by the efficient intramolecular BRET mechanism [77].

Dixon et al. and the authors developed a novel PCA using 3-split NanoLuc [67, 68]. The pair of the small tags is very effective to avoid misfolding and steric hindrance of target proteins. Our next challenge will be to further improve the efficiency and stability of the reconstitution for its wider use.

Acknowledgements

This work was supported partly by the Strategic International Collaborative Research Program (SICORP), Japan Science and Technology Agency, by JSPS KAKENHI, grant numbers JP18H03851 (to HU) and JP17K06920 (to YOM), from the Japan Society for the Promotion of Science, Japan, and by Dynamic Alliance for Open Innovation Bridging Human, Environment and Materials from MEXT, Japan.

Author details

Yuki Ohmuro-Matsuyama and Hiroshi Ueda*
Laboratory for Chemistry and Life Science, Institute of Innovative Research, Tokyo Institute of Technology, Yokohama, Kanagawa, Japan

*Address all correspondence to: ueda@res.titech.ac.jp

References

[1] Stumpf MPH, Thorne T, de Silva E, Stewart R, An HJ, Lappe M, et al. Estimating the size of the human interactome. Proceedings of the National Academy of Sciences of the United States of America. 2008;**105**:6959-6964. DOI: 10.1073/pnas.0708078105

[2] Venkatesan K, Rual JF, Vazquez A, Stelzl U, Lemmens I, Hirozane-Kishikawa T, et al. An empirical framework for binary interactome mapping. Nature Methods. 2009;**6**: 83-90. DOI: 10.1038/nmeth.1280

[3] Zhang QC, Petrey D, Deng L, Qiang L, Shi Y, Thu CA, et al. Structure-based prediction of protein-protein interactions on a genome-wide scale. Nature. 2012;**490**:556-560. DOI: 10.1038/nature11503

[4] Johnsson N, Varshavsky A. Split ubiquitin as a sensor of protein interactions in vivo. Proceedings of the National Academy of Sciences of the United States of America. 1994;**91**:10340-10344

[5] Shekhawat SS, Ghosh I. Split-protein systems: Beyond binary protein-protein interactions. Current Opinion in Chemical Biology. 2011;**15**:789-797. DOI: 10.1016/j.cbpa.2011.10.014

[6] Smirnova DV, Ugarova NN. Firefly luciferase-based fusion proteins and their applications in bioanalysis. Photochemistry and Photobiology. 2017;**93**:436-447. DOI: 10.1111/php.12656

[7] Xing SP, Wallmeroth N, Berendzen KW, Grefen C. Techniques for the analysis of protein-protein interactions in vivo. Plant Physiology. 2016;**171**:727-758. DOI: 10.1104/pp.16.00470

[8] Kerppola TK. Design and implementation of bimolecular fluorescence complementation (BiFC) assays for the visualization of protein interactions in living cells. Nature Protocols. 2006;**1**:1278-1286. DOI: 10.1038/nprot.2006.201

[9] Tchekanda E, Sivanesan D, Michnick SW. An infrared reporter to detect spatiotemporal dynamics of protein-protein interactions. Nature Methods. 2014;**11**:641-644. DOI: 10.1038/Nmeth.2934

[10] To TL, Zhang Q, Shu XK. Structure-guided design of a reversible fluorogenic reporter of protein-protein interactions. Protein Science. 2016;**25**:748-753. DOI: 10.1002/pro.2866

[11] Hall MP, Unch J, Binkowski BF, Valley MP, Butler BL, Wood MG, et al. Engineered luciferase reporter from a deep sea shrimp utilizing a novel imidazopyrazinone substrate. ACS Chemical Biology. 2012;**7**:1848-1857. DOI: 10.1021/cb3002478

[12] England CG, Ehlerding EB, Cai WB. NanoLuc: A small luciferase is brightening up the field of bioluminescence. Bioconjugate Chemistry. 2016;**27**:1175-1187. DOI: 10.1021/acs.bioconjchem.6b00112

[13] Dixon AS, Schwinn MK, Hall MP, Zimmerman K, Otto P, Lubben TH, et al. NanoLuc complementation reporter optimized for accurate measurement of protein interactions in cells. ACS Chemical Biology. 2016;**11**:400-408. DOI: 10.1021/acschembio.5b00753

[14] Verhoef LGGC, Mattioli M, Ricci F, Li YC, Wade M. Multiplex detection of protein-protein interactions using a next generation luciferase reporter. Biochimica et Biophysica Acta-Molecular Cell Research. 2016;**1863**: 284-292. DOI: 10.1016/j.bbamcr.2015.11.031

[15] deMan FHAF, Cabezas MC, vanBarlingen HHJJ, Erkelens DW, deBruin TWA. Triglyceride-rich lipoproteins in non-insulin-dependent diabetes mellitus: Post-prandial metabolism and relation to premature atherosclerosis. European Journal of Clinical Investigation. 1996;**26**:89-108. DOI: 10.1046/j.1365-2362.1996.114256.x

[16] Subramanian S, Chait A. Hypertriglyceridemia secondary to obesity and diabetes. Biochimica Et Biophysica Acta-Molecular and Cell Biology of Lipids. 2012;**1821**:819-825. DOI: 10.1016/j.bbalip.2011.10.003

[17] Do R, Willer CJ, Schmidt EM, Sengupta S, Gao C, Peloso GM, et al. Common variants associated with plasma triglycerides and risk for coronary artery disease. Nature Genetics. 2013;**45**:1345-1352. DOI: 10.1038/ng.2795

[18] Kannel WB, Vasan RS. Triglycerides as vascular risk factors: New epidemiologic insights. Current Opinion in Cardiology. 2009;**24**:345-350. DOI: 10.1097/HCO.0b013e32832c1284

[19] Miller M, Stone NJ, Ballantyne C, Bittner V, Criqui MH, Ginsberg HN, et al. Triglycerides and cardiovascular disease a scientific statement from the American Heart Association. Circulation. 2011;**123**:2292-2333. DOI: 10.1161/CIR.0b013e3182160726

[20] Chi X, Britt EC, Shows HW, Hjelmaas AJ, Shetty SK, Cushing EM, et al. ANGPTL8 promotes the ability of ANGPTL3 to bind and inhibit lipoprotein lipase. Molecular Metabolism. 2017;**6**:1137-1149. DOI: 10.1016/j.molmet.2017.06.014

[21] Garnier M, Zaratin PF, Ficalora G, Valente M, Fontanella L, Rhee MH, et al. Up-regulation of regulator of G protein signaling 4 expression in a model of neuropathic pain and insensitivity to morphine. The Journal of Pharmacology and Experimental Therapeutics. 2003;**304**:1299-1306. DOI: 10.1124/jpet.102.043471

[22] James MA, Lu Y, Liu Y, Vikis HG, You M. RGS17, an overexpressed gene in human lung and prostate cancer, induces tumor cell proliferation through the cyclic AMP-PKA-CREB pathway. Cancer Research. 2009;**69**:2108-2116. DOI: 10.1158/0008-5472.Can-08-3495

[23] Lerner TN, Kreitzer AC. RGS4 is required for dopaminergic control of striatal LTD and susceptibility to Parkinsonian motor deficits. Neuron. 2012;**73**:347-359. DOI: 10.1016/j.neuron.2011.11.015

[24] Sokolov E, Iannitti DA, Schrum LW, McKillop IH. Altered expression and function of regulator of G-protein signaling-17 (RGS17) in hepatocellular carcinoma. Cellular Signalling. 2011;**23**:1603-1610. DOI: 10.1016/j.cellsig.2011.05.012

[25] Li YH, Li LL, Lin JY, Hu X, Li BX, Xue AM, et al. Deregulation of RGS17 expression promotes breast cancer progression. Journal of Cancer. 2015;**6**:767-775. DOI: 10.7150/jca.11833

[26] Bodle CR, Hayes MP, O'Brien JB, Roman DL. Development of a bimolecular luminescence complementation assay for RGS: G protein interactions in cells. Analytical Biochemistry. 2017;**522**:10-17. DOI: 10.1016/j.ab.2017.01.013

[27] Reyes-Alcaraz A, Lee YN, Yun S, Hwang JI, Seong JY. Conformational signatures in beta-arrestin2 reveal natural biased agonism at a G-protein-coupled receptor. Communications Biology. 2018;**1**:128. DOI: 10.1038/s42003-018-0134-3

[28] Storme J, Cannaert A, Van Craenenbroeck K, Stove CP. Molecular dissection of the human A(3) adenosine receptor coupling with beta-arrestin2.

Biochemical Pharmacology. 2018;**148**: 298-307. DOI: 10.1016/j.bcp.2018.01.008

[29] Habara M, Mori N, Okada Y, Kawasumi K, Nakao N, Tanaka Y, et al. Molecular characterization of feline melanocortin 4 receptor and melanocortin 2 receptor accessory protein 2. General and Comparative Endocrinology. 2018;**261**:31-39. DOI: 10.1016/j.ygcen.2018.01.020

[30] Leroy E, Balligand T, Pecquet C, Mouton C, Colau D, Shiau AK, et al. Differential effect of inhibitory strategies of the V617 mutant of JAK2 on cytokine receptor signaling. The Journal of Allergy and Clinical Immunology. 2019 (in press). DOI: 10.1016/j.jaci.2018.12.1023

[31] O'Neill S, Mathis M, Kovacic L, Zhang S, Reinhardt J, Scholz D, et al. Quantitative interaction analysis permits molecular insights into functional NOX4 NADPH oxidase heterodimer assembly. The Journal of Biological Chemistry. 2018;**293**:8750-8760. DOI: 10.1074/jbc.RA117.001045

[32] Chaudhri A, Xiao Y, Klee AN, Wang X, Zhu B, Freeman GJ. PD-L1 binds to B7-1 only in cis on the same cell surface. Cancer Immunology Research. 2018;**6**:921-929. DOI: 10.1158/2326-6066.CIR-17-0316

[33] Halls ML, Bathgate RA, Sutton SW, Dschietzig TB, Summers RJ. International union of basic and clinical pharmacology. XCV. Recent advances in the understanding of the pharmacology and biological roles of relaxin family peptide receptors 1-4, the receptors for relaxin family peptides. Pharmacological Reviews. 2015;**67**:389-440. DOI: 10.1124/pr.114.009472

[34] Bathgate RA, Halls ML, van der Westhuizen ET, Callander GE, Kocan M, Summers RJ. Relaxin family peptides and their receptors. Physiological Reviews. 2013;**93**:405-480. DOI: 10.1152/physrev.00001.2012

[35] Ivell R, Agoulnik AI, Anand-Ivell R. Relaxin-like peptides in male reproduction—A human perspective (vol 174, pg 990, 2017). British Journal of Pharmacology. 2017;**174**:4836-4836

[36] Patil NA, Rosengren KJ, Separovic F, Wade JD, Bathgate RAD, Hossain MA. Relaxin family peptides: Structure-activity relationship studies. British Journal of Pharmacology. 2017;**174**: 950-961. DOI: 10.1111/bph.13684

[37] Hu MJ, Shao XX, Li HZ, Nie WH, Wang JH, Liu YL, et al. Development of a novel ligand binding assay for relaxin family peptide receptor 3 and 4 using NanoLuc complementation. Amino Acids. 2018;**50**:1111-1119. DOI: 10.1007/s00726-018-2588-5

[38] Wang JH, Nie WH, Shao XX, Li HZ, Hu MJ, Liu YL, et al. Exploring electrostatic interactions of relaxin family peptide receptor 3 and 4 with ligands using a NanoBiT-based binding assay. Biochimica et Biophysica Acta - Biomembranes. 2019;**1861**:776-786. DOI: 10.1016/j.bbamem.2019.01.010

[39] Loffler M, Morote-Garcia JC, Eltzschig SA, Coe IR, Eltzschig HK. Physiological roles of vascular nucleoside transporters. Arteriosclerosis, Thrombosis, and Vascular Biology. 2007;**27**:1004-1013. DOI: 10.1161/Atvbaha.106.126714

[40] Ramakers BP, Riksen NP, Stal TH, Heemskerk S, van den Broek P, Peters WHM, et al. Dipyridamole augments the antiinflammatory response during human endotoxemia. Critical Care. 2011;**15**:R289. DOI: 10.1186/cc10576

[41] Elsherbiny NM, Al-Gayyar MM, Abd El Galil KH. Nephroprotective role of dipyridamole in diabetic nephropathy: Effect on inflammation

and apoptosis. Life Sciences. 2015;**143**: 8-17. DOI: 10.1016/j.lfs.2015.10.026

[42] Grañe-Boladeras N, Williams D, Tarmakova Z, Stevanovic K, Villani LA, Mehrabi P, et al. Oligomerization of equilibrative nucleoside transporters: A novel regulatory and functional mechanism involving PKC and PP1. The FASEB Journal. 2019;**33**:3841-3850. DOI: 10.1096/fj.201800440RR

[43] Cannaert A, Storme J, Franz F, Auwarter V, Stove CP. Detection and activity profiling of synthetic cannabinoids and their metabolites with a newly developed bioassay. Analytical Chemistry. 2016;**88**:11476-11485. DOI: 10.1021/acs.analchem.6b02600

[44] Cannaert A, Franz F, Auwarter V, Stove CP. Activity-based detection of consumption of synthetic cannabinoids in authentic urine samples using a stable cannabinoid reporter system. Analytical Chemistry. 2017;**89**:9527-9536. DOI: 10.1021/acs.analchem.7b02552

[45] Cannaert A, Vasudevan L, Friscia M, Mohr ALA, Wille SMR, Stove CP. Activity-based concept to screen biological matrices for opiates and (synthetic) opioids. Clinical Chemistry. 2018;**64**:1221-1229. DOI: 10.1373/clinchem.2018.289496

[46] Arai T, Hasegawa M, Akiyama H, Ikeda K, Nonaka T, Mori H, et al. TDP-43 is a component of ubiquitin-positive tau-negative inclusions in frontotemporal lobar degeneration and amyotrophic lateral sclerosis. Biochemical and Biophysical Research Communications. 2006;**351**:602-611. DOI: 10.1016/j.bbrc.2006.10.093

[47] Neumann M, Sampathu DM, Kwong LK, Truax AC, Micsenyi MC, Chou TT, et al. Ubiquitinated TDP-43 in frontotemporal lobar degeneration and amyotrophic lateral sclerosis. Science. 2006;**314**:130-133. DOI: 10.1126/science.1134108

[48] Oberstadt M, Stieler J, Simpong DL, Romuss U, Urban N, Schaefer M, et al. TDP-43 self-interaction is modulated by redox-active compounds Auranofin, Chelerythrine and Riluzole. Scientific Reports. 2018;**8**:2248. DOI: 10.1038/s41598-018-20565-0

[49] Nguyen GK, Kam A, Loo S, Jansson AE, Pan LX, Tam JP. Butelase 1: A versatile ligase for peptide and protein macrocyclization. Journal of the American Chemical Society. 2015;**137**:15398-15401. DOI: 10.1021/jacs.5b11014

[50] Antos JM, Truttmann MC, Ploegh HL. Recent advances in sortase-catalyzed ligation methodology. Current Opinion in Structural Biology. 2016;**38**:111-118. DOI: 10.1016/j.sbi.2016.05.021

[51] Fierer JO, Veggiani G, Howarth M. SpyLigase peptide-peptide ligation polymerizes affibodies to enhance magnetic cancer cell capture. Proceedings of the National Academy of Sciences of the United States of America. 2014;**111**:E1176-E1181. DOI: 10.1073/pnas.1315776111

[52] Siegmund V, Piater B, Zakeri B, Eichhorn T, Fischer F, Deutsch C, et al. Spontaneous isopeptide bond formation as a powerful tool for engineering site-specific antibody-drug conjugates. Scientific Reports. 2016;**6**:39291. DOI: 10.1038/srep39291

[53] Li N, Yu Z, Ji Q, Sun JY, Liu X, Du MJ, et al. An enzyme-mediated protein-fragment complementation assay for substrate screening of sortase A. Biochemical and Biophysical Research Communications. 2017;**486**:257-263. DOI: 10.1016/j.bbrc.2017.03.016

[54] Zhao J, Nelson TJ, Vu Q, Truong T, Stains CI. Self-assembling NanoLuc luciferase fragments as probes for protein aggregation in living cells. ACS Chemical Biology. 2016;**11**:132-138. DOI: 10.1021/acschembio.5b00758

[55] Fuentealba RA, Marasa J, Diamond MI, Piwnica-Worms D, Weihl CC. An aggregation sensing reporter identifies leflunomide and teriflunomide as polyglutamine aggregate inhibitors. Human Molecular Genetics. 2012;**21**: 664-680. DOI: 10.1093/hmg/ddr500

[56] Chun WJ, Waldo GS, Johnson GVW. Split GFP complementation assay: A novel approach to quantitatively measure aggregation of tau in situ: Effects of GSK3 beta activation and caspase 3 cleavage. Journal of Neurochemistry. 2007;**103**:2529-2539. DOI: 10.1111/j.1471-4159.2007.04941.x

[57] Cabantous S, Terwilliger TC, Waldo GS. Protein tagging and detection with engineered self-assembling fragments of green fluorescent protein. Nature Biotechnology. 2005;**23**:102-107. DOI: 10.1038/nbt1044

[58] Cabantous S, Waldo GS. In vivo and in vitro protein solubility assays using split GFP. Nature Methods. 2006;**3**: 845-854. DOI: 10.1038/nmeth932

[59] Hashimoto T, Adams KW, Fan Z, McLean PJ, Hyman BT. Characterization of oligomer formation of amyloid-beta peptide using a split-luciferase complementation assay. The Journal of Biological Chemistry. 2011;**286**:27081-27091. DOI: 10.1074/jbc.M111.257378

[60] Listwan P, Terwilliger TC, Waldo GS. Automated, high-throughput platform for protein solubility screening using a split-GFP system. Journal of Structural and Functional Genomics. 2009;**10**: 47-55. DOI: 10.1007/s10969-008-9049-4

[61] Leonetti MD, Sekine S, Kamiyama D, Weissman JS, Huang B. A scalable strategy for high-throughput GFP tagging of endogenous human proteins. Proceedings of the National Academy of Sciences of the United States of America. 2016;**113**:E3501-E3508. DOI: 10.1073/pnas.1606731113

[62] Schwinn MK, Machleidt T, Zimmerman K, Eggers CT, Dixon AS, Hurst R, et al. CRISPR-mediated tagging of endogenous proteins with a luminescent peptide. ACS Chemical Biology. 2018;**13**:467-474. DOI: 10.1021/acschembio.7b00549

[63] Oh-Hashi K, Furuta E, Fujimura K, Hirata Y. Application of a novel HiBiT peptide tag for monitoring ATF4 protein expression in Neuro2a cells. Biochemistry and Biophysics Reports. 2017;**12**:40-45. DOI: 10.1016/j.bbrep.2017.08.002

[64] Sasaki M, Anindita PD, Phongphaew W, Carr M, Kobayashi S, Orba Y, et al. Development of a rapid and quantitative method for the analysis of viral entry and release using a NanoLuc luciferase complementation assay. Virus Research. 2018;**243**:69-74. DOI: 10.1016/j.virusres.2017.10.015

[65] Tamura T, Fukuhara T, Uchida T, Ono C, Mori H, Sato A, et al. Characterization of recombinant *Flaviviridae* viruses possessing a small reporter tag. Journal of Virology. 2018;**92**:e01582-17. DOI: 10.1128/JVI.01582-17

[66] Cabantous S, Nguyen HB, Pedelacq JD, Koraichi F, Chaudhary A, Ganguly K, et al. A new protein-protein interaction sensor based on tripartite split-GFP association. Scientific Reports. 2013;**3**:2854. DOI: 10.1038/srep02854

[67] Dixon AS, Kim SJ, Baumgartner BK, Krippner S, Owen SC. A tri-part protein complementation system using AntibodySmall peptide fusions enables homogeneous immunoassays. Scientific Reports. 2017;**7**:8186. DOI: 10.1038/s41598-017-07569-y

[68] Ohmuro-Matsuyama Y, Ueda H. Homogeneous noncompetitive luminescent immunodetection of small molecules by ternary protein

fragment complementation. Analytical Chemistry. 2018;**90**:3001-3004. DOI: 10.1021/acs.analchem.7b05140

[69] Ueda H, Tsumoto K, Kubota K, Suzuki E, Nagamune T, Nishimura H, et al. Open sandwich ELISA: A novel immunoassay based on the interchain interaction of antibody variable region. Nature Biotechnology. 1996;**14**:1714-1718. DOI: DOI 10.1038/nbt1296-1714

[70] Thorne N, Auld DS, Inglese J. Apparent activity in high-throughput screening: Origins of compound-dependent assay interference. Current Opinion in Chemical Biology. 2010;**14**:315-324. DOI: 10.1016/j.cbpa.2010.03.020

[71] Thorne N, Shen M, Lea WA, Simeonov A, Lovell S, Auld DS, et al. Firefly luciferase in chemical biology: A compendium of inhibitors, mechanistic evaluation of chemotypes, and suggested use As a reporter. Chemistry & Biology. 2012;**19**:1060-1072. DOI: 10.1016/j.chembiol.2012.07.015

[72] Auld DS, Southall NT, Jadhav A, Johnson RL, Diller DJ, Simeonov A, et al. Characterization of chemical libraries for luciferase inhibitory activity. Journal of Medicinal Chemistry. 2008;**51**:2372-2386. DOI: 10.1021/jm701302v

[73] Auld DS, Narahari J, Ho PI, Casalena D, Nguyen V, Cirbaite E, et al. Characterization and use of TurboLuc luciferase as a reporter for high-throughput assays. Biochemistry. 2018;**57**:4700-4706. DOI: 10.1021/acs.biochem.8b00290

[74] Ho PI, Yue K, Pandey P, Breault L, Harbinski F, McBride AJ, et al. Reporter enzyme inhibitor study to aid assembly of orthogonal reporter gene assays. ACS Chemical Biology. 2013;**8**:1009-1017. DOI: 10.1021/cb3007264

[75] Dale R, Ohmuro-Matsuyama Y, Ueda H, Kato N. Mathematical model of the firefly luciferase complementation assay reveals a non-linear relationship between the detected luminescence and the affinity of the protein pair being analyzed. PLoS One. 2016;**11**:e0148256. DOI: 10.1371/journal.pone.0148256

[76] Walker JR, Hall MP, Zimprich CA, Robers MB, Duellman SJ, Machleidt T, et al. Highly potent cell-permeable and impermeable NanoLuc luciferase inhibitors. ACS Chemical Biology. 2017;**12**:1028-1037. DOI: 10.1021/acschembio.6b01129

[77] Suzuki K, Kimura T, Shinoda H, Bai GR, Daniels MJ, Arai Y, et al. Five colour variants of bright luminescent protein for real-time multicolour bioimaging. Nature Communications. 2016;**7**:13718. DOI: 10.1038/ncomms13718

Chapter 3

Biosensors Using Free and Immobilized Cells of Luminous Bacteria

Anvar D. Ismailov, Leyla E. Aleskerova, Kristina A. Alenina and Elena N. Efremenko

Abstract

The technologies of receiving free and immobilized photobacteria cells for biomonitoring of toxins are considered. The mechanisms of interaction of toxins with photobacteria are observed. The main attention is paid to the immobilized procedures and structures of carriers. Data on poly(vinyl)alcohol (PVA) cryogel immobilization of different strains of photobacteria are presented. It is established that intensity and stability of light emission of PVA cells is competently controlled by: (1) intensity and persistence of a luminescent cycle using bacterial strain; (2) type of the carrier and the composition of the gel-formation medium; (3) freeze-thawing procedures; and (4) physical and chemical conditions of storage and application. The developed technology of cryogenic gel formation has kept the survival of luminous bacteria in the carrier practically at 100% without the introduction of additional cryoprotecting agents and procedures of a light induction. With storage at −80°C, bioluminescent activity remained without changes about 2 years. Using the immobilized preparations of biosensor, the discrete and continuous analysis of heavy metals, chlorophenols, and pesticides is carried out. The sensitivity of free and immobilized cells to the chosen toxicants is approximately identical. The continuous monitoring of toxicant conditions is optimized.

Keywords: photobacteria, biosensors, immobilization, poly(vinyl)alcohol, biomonitoring

1. Toxin action on the photobacteria light emission

The interaction mechanism was investigated only for certain groups of chemicals and mainly with the use of free cells. Toxic agents that suppressed the emission of photobacteria can be divided into classes rather conditionally.

- On the chemical structure: heavy metals, electron acceptors, respiratory poisons, aliphatic, aromatic and heterocyclic hydrocarbons, alcohols, ketones, acids, and others. The level of toxin hydrophobicity is important.

- On the type of targets: membrane active substances, specific inhibitors of the genetic apparatus, and inhibitors of the energy and lipid metabolism enzymes.

It well known also the specific inhibitors of luciferase and the bioluminescent enzyme system, and auxiliary systems for the biosynthesis of substrates and regulatory molecules.

These substances suppress the light emission and some can cause small activation of the luminescence of bacteria. Naturally, the determining factor of interaction with the cell is the chemical structure of the substance, concentration, and time of incubation. These parameters are manifested in the kinetics of changes in luminescence and the magnitude of the inhibition constant. The main options in the kinetics of inhibition are presented in **Figure 1**. Inhibition can occur almost instantaneously (Curve 1) or with a lag phase, depending on the nature and concentration of the substance (Curve 3). In some cases, the effect may be reversible (Curve 2).

The mechanisms of interaction of toxins with photobacteria are observed. Data on poly(vinyl)alcohol (PVA) cryogel immobilization of different strains of photobacteria are presented. It is established that intensity and stability of light emission of PVA cells are competently controlled by: (1) intensity and persistence of a luminescent cycle using bacterial strain; (2) type of the carrier and the composition of the gel-formation medium; (3) freeze-thawing procedures; and (4) physical and chemical conditions of storage and application. The developed technology of cryogenic gel formation has kept the survival of luminous bacteria in the carrier practically at 100% without introduction of additional cryoprotecting agents and procedures of a light induction. Specific bioluminescent activity was restored to level of activity of free cells. At storage at −80°C, bioluminescent activity remained without changes about 2 years. The detecting level of a light emission at 4°C for psychrophilic strains over 1 month, at 20°C for 3 days. High survival of PVA-immobilized cells and prolonged light emission reflect advantages of cryogenic immobilization of photobacteria. With the use of immobilized preparations of biosensor, the discrete and continuous analysis of heavy metals, chlorophenols, and pesticides is carried out. The sensitivity of free and immobilized cells to the chosen toxicants is approximately identical. The continuous monitoring of toxicant conditions is optimized. The analysis using photo-PVA-biosensors can be perspective

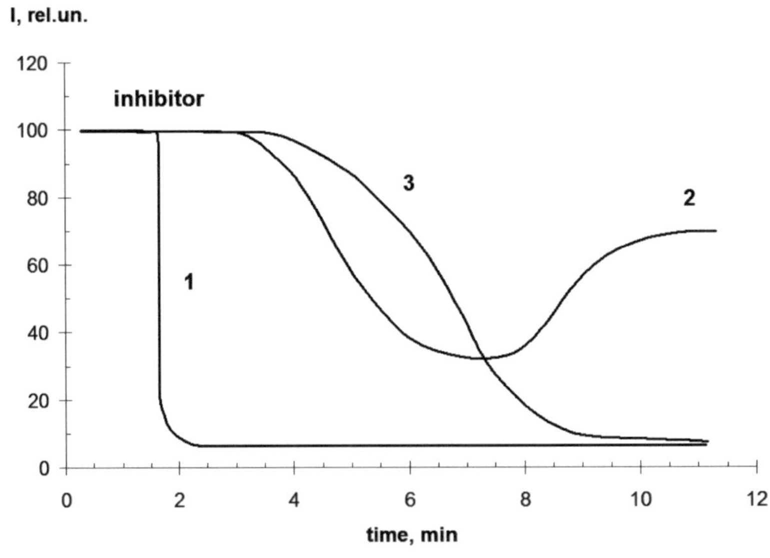

Figure 1.
The kinetics of the Vibrio harveyi free cell luminescence with various chemicals. 1–instant inhibition; 2–reversible inhibition; 3–lag-phase inhibition. Incubation medium: 3% NaCl +0.1 M phosphate buffer, pH 7.0.

for express monitoring of environments in a mode of real time in water areas and industrial wastes, and also in the control over processes of biotransformation and degradation of toxins.

Irreversible inhibition causes a wide class of toxins of organic and inorganic nature; reversible inhibition is observed under the action of electron acceptors, which restore and lose their inhibitory properties during incubation; and inhibition with lag-phase implies sequential interaction with membrane and intracellular structures, including a diffusion step.

2. Photo-biosensors

Photobacteria, as well as other microorganisms with cloned operon of luminescent system of photobacteria, and recombinant cells with luminescent activity are widely used in various biotechnological tasks, and, first of all, in ecological biomonitoring of toxins and other biologically active agents. Dates are summarized in reviews [1, 2]. Depending on the purposes of researches in technological procedures, both the free and immobilized cells are used [3, 4].

The use of preparations with immobilized cells leads to heterogeneous catalysis that in many cases provides economic and practical efficiency. The advantages of this approach are primarily related to simplification of the manufacturing process as well as the opportunity to switch from occasional output of desired products or expected substrate reactions toward a continuous process. In addition, immobilized cells allow using preparations multiple times and for longer period compared to single application of cells in a homogeneous suspension. The technology of immobilization in many cases enhances enzymatic activity. Also, immobilized preparations have increased resistance to various physicochemical factors (temperature and pH) and lesser sensitivity to the action of pathogenic organisms. The results of a 10-year long examination evaluating technologies creating genetically engineered microbial strains as well as immobilization procedures, storage, and application of genetically engineered microbial strains in both laboratory and practical application of biosensors in toxicological biomonitoring were thoroughly described [1, 2, 5].

Immobilizing photobacteria increases stability of a biosensor upon its storage as well as duration of its use [6–8]. Similarly to the free cells in suspension, in the case of immobilized preparations, an integral toxicity assay is performed by natural photobacteria, whereas specific toxin detection is mediated by mutant and genetically engineered microbial strains containing cloned genes of a bioluminescent system. Moreover, it should be mentioned that during a specific assay with genetically engineered microbial strains, light emission is observed after a lag-phase lasting for many hours. Various reporter bacteria are used for specific analysis of various substances [9]. A reporter microbial culture was immobilized and fixed on the surface of a photodetector element using Sr-alginate gel. Immobilized biosensors for detecting such substances were used for both during discrete and continuous monitoring. It was found that both the maximum emission and rate of emission induction depended on incubation time and concentration of substances. It was concluded that this system was suitable for real-time analysis of environmental pollutants and industrial waste as well as evaluation of metabolic status of various biological materials. A linear dependence between induced bioluminescent response and concentration of the substances was noted using both free and immobilized bacteria. The authors attracted attention to high stability of the alginate-glycerol cell suspension at −70°C. The data presented in [7] were obtained using recombinant Salmonella typhimurium strain TA1535 designed from a plasmid with an inducible SOS-promoter linked to the luxCDABE operon derived from *Photobacterium leiognathi*. The mixture was immobilized in agar carrier, a simple procedure that includes a plate matrix (10 μl of mixture added into the wells) incubated for 30 min for polymerization and further storage at 4°C. The principal protocol used for analysis is similar to others that use recombinant biosensors and includes emission induction upon incubation with toxic substance. Model experiments on light induction with this bacterial strain were made with DNA-damaging agent—mitomycin C. The immobilized cells retained activity at 4°C for 6 weeks. A response of 4-week old bacterial culture exposed to mitomycin C was indistinguishable from that triggered by a fresh immobilized culture. Both the magnitude of the maximum response and duration of the lag-phase depend on the concentration of mitomycin C. The maximum activation response is obtained at approximately 1 μg/ml concentration of mitomycin C. It is crucial that further increase in concentrations results in abrupt quenching of the emission. Apparently, this is due to a cytotoxic action of the antibiotic on the bacteria, similar to toxins affecting recombinant bacteria of other strains treated with high concentrations. Bacterial biosensors immobilized in thin cellulose-agar films were used to detect various chemicals such as phenol, hydrogen peroxide, copper, and cadmium [10]. It was found that immobilized cells revealed high sensitivity to these agents, and inhibitory effect was observed at concentrations close to those detected by free bacteria. The preparation was stable when dissolved in AWS culture medium for 4 weeks at 4°C. However, incubation in 3% NaCl solution significantly decreased stability. Along with toxicology experiments, data regarding temperature and pH, dependent relations with luminescence of both free and immobilized bacterial preparations, as well as spectral and kinetic characteristics are presented. A number of studies were done to create novel-type biosensors to be used in environmental toxicological analysis [11, 12]. A fundamental difference for such systems was that they contained a combined complex made of optical fiber detector together with a biodetector. Results on improving technology of generating combined biosensors containing natural and recombinant microbial strains immobilized on optical fiber strands and used for cytotoxicity and genotoxicity assays are presented in [13].

Combined systems are considered especially promising for practical application of immobilized cells of type "biosensor-photoconductor l-photodetector." Many of these were aimed at developing genetically engineered constructions, containing sensor genes: SOS-system, heat shock defense system, DNA, and membrane-damaging agents, specific to various toxins, and reporter genes derived from photobacteria. All strains were immobilized at the different optic fiber strands. Therein, Ca-alginate gel was used as a carrier. In such systems, each biosensor specifically reacted to a certain type of toxins. The developed test system was effective in detecting pure inorganic and organic toxins as well as admixtures of toxic substances in aqueous medium, soil, and other samples. Natural photobacteria were used together with genetically engineered counterparts, where toxic agents had an inhibitory effect on their luminescence. Application of such bacteria based on a need to have a reference control of the cytotoxic action. It was found that thin films consisting of gel and cells were stable for 6 h at 26°C and able to detect mitomycin C at concentrations up to 25 μg/l. Concentrations of cells up to $(1–3) \times 10^7$ in a carrier were effective for different analytical procedures. Biosensor stability within a combined system sensitive to temperature impact, as well as other chemical and physical parameters of the bio-detecting system, was analyzed in detail in response to substances causing a stressful reaction in the cells (heat shock, action of SOS agents, impaired protein biosynthesis, peroxide, and oxidative stress).

However, it should be noted that cells immobilized in agar, agarose, and alginate have insufficient stability of luminescence, due to the sensitivity of the cells to relatively high (up to 30–50°C) gelation temperature. It is worth mentioning that alginate gels with relatively low gelation temperature can be used as well. Nonetheless, the effect of temperature results in the development of dark mutants of photobacteria having a temperature optimum for luminescence at 15–25°C (depending on species). The salt composition contained in the culture media plays an important role for marine photobacteria. Optimal concentrations for emission activity of cells are reached in 2–6% NaCl solution. Ca-alginate gels can be somewhat destabilized in such high salt concentration solutions due to the process when Ca^{2+} ions derived from the gel are replaced by Na^{+} ions. Moreover, complex formation of Ca^{2+} and Sr^{2+} with phosphate and carbonate ions used as buffer solutions may occur.

The data present in [4, 14, 15] the immobilization of bacteria *Photobacterium phosphoreum* in different gel formation materials: agar, agarose, carrageenan, polyacrylamide, calcium alginate, polyvinyl alcohol, polyurethane, calcium carboxymethyl cellulose, etc. It is established that the used carriers and technologies of an immobilization extremely and variously influence the stability of photobacteria. The duration of a luminescence alginate-glycerol suspension of cells, at incubation in 3% NaCl, at 4°C, reached 4 weeks (the minimal detected luminescence level—up to 6 weeks). The duration of a luminescence of free cells in the same conditions did not exceed 2 weeks. In some cases the immobilized cells in an agar have significantly smaller stability, than that of free cells. The agarose-immobilized cells possess approximately the same duration of luminescence, as free cells—2 weeks. The alginate glycerol suspension completely kept the initial level of luminescence at storage at −80°C within 12 weeks. In storage, at –20°C, emission activity decreased, similarly observed at 4°C.

The investigations [3, 16, 17] are directed to the development of technologies of bioluminescent monitoring of chemicals in the discrete and continuous mode of bio-testing. The stated explanation follows from the well-known temperature effects on a luminescence of photobacteria. Temperature influence (30–36°C, 30–60 min) leads to the formation of dim and dark mutants [18]. The temperature factor makes especially strong effect on *Photobacterium phosphoreum* bacteria, which has a maximum bioluminescent activity at 15–18°C. Data on luminescence kinetics are important during an incubation of preparations at positive temperature. The unstability of luminescence that is caused with physical states of using gels was established. The decreasing of light emission caused also with partial replacement of ions of Ca^{2+} in the carrier on Na^{+} ions. Replacement of Ca^{2+} on Sr^{2+}-ions in the carrier increased its stability. The analytical procedure using Ca and Sr-alginate gels is approved on five heavy metals: $Pb(NO_3)_2$, $NaAsO_2$, $NiCl_2$, $CdCl_2$, $HgCl_2$, and also SDS and pentachlorophenol (PCP). The authors note essential distinctions in inhibition kinetics of the specified toxins between the free and immobilized cells, but the sensitivity is rather close. Besides them, stimulation of a luminescence by low concentration of PCP (less than 0.1 ppm) and $CdCl_2$ (less than 10 ppm) was observed. The procedures developed for discrete biomonitoring have been used for the continuous analysis of toxins. The interaction of toxins with cellular targets depends on hydrodynamic parameters. The optimized conditions were: speed of a channel is 25 ml an hour, delay time after introduction of toxin of 37 s, time interaction 1–2 min, and cooling system has been used. Stability of a luminescence is more than 40 min. Pulse introduction of a toxin to a channel with the immobilized cells leads to reversible kinetics of inhibition. At washing away of a toxin by the stream observes complete or partial recovery of emission, allows to repeated biomonitoring on the same biosensor. The effect "dose/time" is shown in a kinetic profile of suppression and reversion of a luminescence [16]. The fact that the cells of *Photobacterium phosphoreum* immobilized in Sr-alginate gel can effectively be used for discrete and continuous detection of toxins can be considered as the main conclusion of this work.

The work [17] was done for biomonitoring of water toxicity based on continuous cultivation of *Photobacterium phosphoreum*. Attention in this work is paid to a possible problem of emergence of dark mutants (after 10 days of incubation), which begin to dominate, causing suppression of a luminescence. For an exception of this problem, the authors have offered the system based on the special fluidized-bed reactor. The cell-alginate suspension continuously moved in a stream. It is established that domination of dark forms significantly stops in the internal volume of carrier, and suppression of a luminescence is not observed.

Rather high speed of dilution prevents settling by other microorganisms of the reactor. Concentration of cells and emission activity in the developed mode of cultivation is sufficient for observation of the water environment toxicity within 4 weeks. Essentially that photobacteria are capable to grow in granules and remain in water. It is noted that after 24 h in all volume of granules microcolonies of various forms were formed. The survival of cells in the carrier during the procedure of an immobilization, storage, and use of the analytical system was studied. The assessment was carried out by calculation of colonies after destruction of a carrier by $Na_6O_{18}P_6$.

It has been established that all cells are distributed in volume of a matrix; the insignificant quantity is connected at the surface of the carrier. In the developed cultivation mode, 80% of cells remain in the wild type within 40 h. The dark mutants were formed on the seventh day, and then increased up to 100% by the tenth day of cultivation. The authors have made the assumption that mutations mention all cells—immobilized and leaved matrix cells. However, these distinctions are insignificant. It is suggested that matrix protects the immobilized cells from a mutagenesis. Results are considered as a possibility to use of system for the analysis of water toxicity in real time. At the same time, the developed technological operations significantly protect from a dark mutagenesis. The use of analytical system, in this case, is possible within more than 30 days. The measures of protection against infection with other microorganisms are enclosed. The model experiments on the analysis of $HgCl_2$ are given using immobilized and free cells. It is established that the immobilized cells are more sensitive to these chemicals, since free cells are more active it results in more resistance, which is toxic. Cells in the immobilized material react to existence in the $HgCl_2$ up to 0.5 mg/l.

The new technology solution of complex biomonitoring of toxins is submitted in work [19]. The test system based on the multi-channel apparatus for detecting the water environments using some recombinant strains containing lux-CDABE operon is developed.

Each channel of the system was designed from two mini-bioreactors necessary for step continuous process. Each channel contained a certain recombinant strain with the cloned gene of photobacteria: DPD2440-(fabA:luxCDABE), DPD2794-(recA:luxCDBE), and TV1061-(grpE:luxCDABE). The strains are sensitive to the membrane, DNA, and protein destroying agents, respectively. Interaction of these agents with recombinant bacteria causes induction of a luminescence. As a control for cellular toxicity, wild strain, whose bioluminescent reaction was suppressed with toxins, has been used. Some experiments are executed with phenol and mitomycin C. The procedures of biomonitoring for the subsequent applications in practical tasks of the analysis of toxic chemicals of the environment are optimized. For the practical analysis, tests of water from two different (nuclear and thermo-electronic) reactors were used.

Continuous biomonitoring was carried out by cultivation of bacteria in two reactors; the biomass of bacteria was grown in the first cultivator and pumped at the second reactor with toxin injected. Each channel reflected the specific profile of the bioluminescence, corresponding to the chemical nature of the toxic agent. Comparison of a bioluminescent signal between standard toxic substance and the following water solution allowed to define the true equivalent toxicity in a water stream.

The developed system of continuous multi-channel biomonitoring of toxins can be considered, according to authors, as the new strategy of protection of bio-objects against environment pollutants (alternative system of express monitoring and control of the water environment).

Special attention is paid to receiving recombinant strains, and also technological operations of biomonitoring in a channel that are fulfilled, optimized on models, and are applied in water analysis to cool the nuclear and thermo-installation.

The authors note that the tests used for biomonitoring are stable within 1 week at 4°C and pH 7.0. Test-system was carried out in real-time monitoring with the use of automatic computer control.

All used chemicals stimulate the luminescence, however for all the difference in the intensity of light and duration was observed. The induction time is rather long, up to 500 min. At the same time, the high concentration of toxins not only caused induction but also suppression of the luminescence of recombinant strains. It specifies that at high concentration there is no specificity of reaction of recombinant strains. Metabolism of all recombinant strains reacts to high concentration of toxins by suppression of luminescence. It was shown that at certain concentration, the genotoxic action is blocked by cytotoxic action. For the recombinant strains, there are no quantitative data on inhibitory actions on external targets of membranes. With these data, the bioluminescent intensity could be significantly corrected. The results suggest that the water used for the nuclear reactor contains substances with membrane action. It should be noted that the developed system of biomonitoring is rather difficult, consuming, and long, though it, in principle, yields positive results of bio-testing.

The fact that the toxic effect on inhibition of a luminescence is shown on all recombinant strains is important. The developed system of continuous biomonitoring can be considered as the new direction in identification of the nature and toxic action of chemicals and can be also applied to the detection of many industrial wastes. Before the analysis the testified probes have to undergo preliminary processing, for removal of the accompanying bacteria, with the subsequent filtration. The main result of this work is the use of various types of recombinant strains for specific biomonitoring of toxins.

The integrated system of biomonitoring of water (water toxicology) using the test object recombinant bacteria is presented in work [20]. The developed system included four channels with two mini-bioreactors. The recombinant strains of *E. coli*, EBHJ2, DP1, DK1, and DPD279, were used. The chosen strains carry out three specific tasks of cell protection against oxygen shock of superoxide, hydrogen peroxide, and the DNA-damaging agents also. Based on the specified strains, systems are applied for both discrete and continuous monitoring of toxins. The first reactor was used for cultivating the culture of the specified strains with their subsequent transfer in the second bioreactor. It is established that induction of the bioluminescent increase is observed only in the small range of the increasing concentration of the specified inductors. The peroxide and mitomycin, were entered into the second cultivator. The intensity and rate of increase of luminescence are specific to each inductor; and at high concentration, the effect of induction is changed with inhibitory effect.

Particular advantage of the developed system is miniaturization: a little (1–2 ml) reactor volume. The authors assume that the four-channel mini-monitoring system with different recombinant strains is extremely useful viewpoint for the specific analysis of certain classes of toxins in the environment.

3. Immobilization of microorganisms

As carriers for an immobilization of photobacteria, both natural and synthetic materials are used [2, 8]. The immobilization of cells of microorganisms can pass by fixing on a surface and in volume (depending on the used material), or chemical binding on surface or embedding in matrix interior. In most cases, agar, agarose, and alginate gels are used as the carriers for immobilized luminescent bacteria. Ca^{2+} and Sr^{2+} alginate gels are the carriers most commonly used in manufacturing bioluminescent toxicological biosensors.

Methods of an immobilization are various; however, all are directed at increasing the profitability of process and its efficiency. At the same time, special attention is paid to durability for fixing of cells with materials and for prevention of an uncontrolled exit from a carrier phase in incubatory solution. The factors capable to reduce, suppressing metabolic activity, up to violation of integrity of structure of an organism has been studied. First of all, it belongs to temperature, ionic structure, pH, to availability of inhibitors, toxins, and other xenobiotics. Viability and light activity of photobacteria are extremely sensitive to these factors.

At inclusion of cells of microorganisms in carrier volume, a number of factors take part in keeping of cells: mechanical keeping, adsorption, structure of a grid of polymer and the pores, chemical bonds between functional groups of a cellular surface and free groups of the carrier, electrostatic forces, and hydrophobic interactions.

At the same time, the size and structure of pores in the carrier, time diffusion of nutrients or other biologically active agents, and withdrawal of products are important.

4. The immobilization on a surface of the carrier

The immobilization of bacteria in some cases is carried out on a surface of the carrier. It depends on the restrictions for penetration of cells into volume and on the structures of the materials used. Different chemical elements participate in the fixing of cells on the surface of the carrier. Non-covalent interaction is carried out at the contact of cells with the surface of carrier, with the participation of ionic interactions, hydrogen bonds, and hydrophobic forces. In this case, nonspecific cell fixing occurs. When the carrier contains the ligands capable to form rather strong binding with functional groups of a cellular surface, bio-specific binding

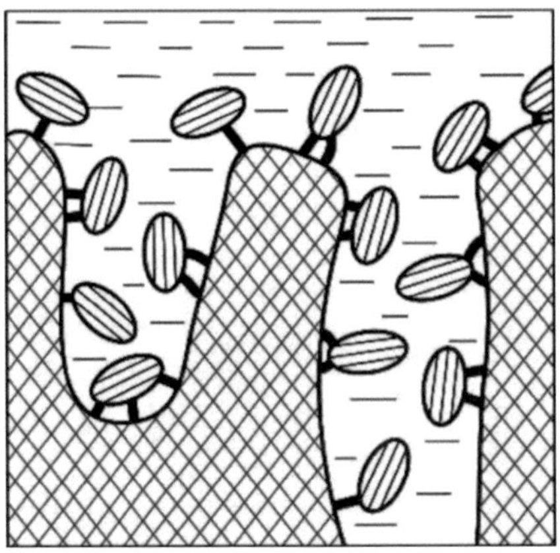

Figure 2.
Cell immobilization on the surface of carrier [8].

is observed. In certain cases for increase in efficiency of process, the carrier has to have special chemical residues capable to form strong bonds with a cellular surface or the so-called sewing elements are used (**Figure 2**).

5. The immobilization in the interior of the carrier

The immobilization in volume of the matrix allows to increase a specific concentration of the cells in comparison with liquid culture, which causes the productivity of biotechnological process. Especially, a possibility of repeated or continuous use of one type carrier with the specific microorganisms should be noted. At the same time, the form, any convenient for operation, can be given to the carrier: granules, plates, threads, tubes, etc. Finally, procedures of biomonitoring become simpler (**Figure 3**).

The procedure of inclusion of cells in the interior of the matrix is generally carried out during incubation of the carrier with biomass, swelling, with the subsequent formation of gels system, with microorganisms, stabilizers, substrates and protectors. Water, through which there is a metabolism, is a part of hydro-gels. For an immobilization of cells in volume of the carrier, a wide range of various materials capable to transition from liquid to "gelatinous," gel/cell composition, under specific conditions is used. The greatest distribution for an immobilization of microorganisms was gained by gelatin, agar, agarose, alginate, carrageenan, etc., in which the spatial grid is supported using non-covalent bands. The specific property of these gels is the transition from liquid state to elastic form, at change of temperature. The majority will be polymerized at heating. However, there are substances that are polymerized with cooling. It is possible to give poly(vinyl) alcohol (PVA) as an example capable to form cryogenic gels. Ion-tropical gels are formed with bands between poly-electrolytic macromolecules.

In polymerization of polysaccharides, the main role in matrix formation is assigned to hydrogen bonds and electrostatic forces. The contribution is made by mechanical packing of polymeric chains in supramolecular structures. Gel-forming properties can differ depending on the source and method of receiving. At the low temperature, the macromolecules can form supramolecular structures with the participation of hydrogen bonds.

The main advantage of the technological procedure is simplicity of inclusion of cells in the carrier without essential violation of structural characteristics, the high catalytic properties [8].

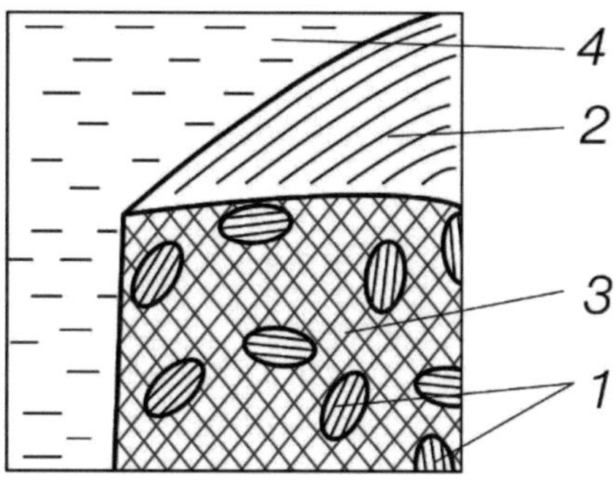

Figure 3.
Immobilization of the cells in the volume of the carrier. 1—Cells, 2—Matrix, 3—Gel-cell mixture, and 4—Carrier [8].

6. Poly(vinyl) alcohol as a carrier for immobilization

Among synthetic polymers, for immobilization of the most different groups of microorganisms, cryogels of poly(vinyl) alcohol (PVA) are used. Because of freezing, elastic hydro-gels are formed. The formation of matrix was participated by hydrogen bands between various sites of polymeric chains. One of the important elements of formation of PVA cryogel is the content of acyl-residues. At 5%, the friable structures were formed. From high concentration of the PVA water solutions (more than 7%), strong structures are formed.

Cryogenic impact on the PVA-water system gives the chance to change properties of polymers and to alter their structure. Stability of the carrier at high negative temperatures is one of the major factors for practical use.

In compliance with literary data, PVA was effectively used in immobilization of many strains of microorganisms. In work [21], different types of microorganisms, which have been successfully immobilized in PVA-cryogel, are summarized.

PVA cryogels have a number of structural and chemical properties, in comparison with other carriers, essential for immobilization of luminous bacteria [22, 23].

First of all, it is connected with availability and low cost of materials, simplicity of the procedure of cryogel formation, and biocompatibility with various types of cells. High resistance of PVA matrix to bio-damage by microorganisms is also important. Essentially, PVA cryogel is not toxic in relation to the included microorganisms. Cryogels have diffusive properties, optimum for detection processes. The electronic microscopy of preparation demonstrates existence of high porosity of the matrix containing micro (less than 1 micron) and macro (more than 1 micron) pores. Existence of a time in many cases lifts a diffusive limit for chemicals (substrata, salts, and products of metabolism of cells), and also activators and inhibitors of this or that of processes. Physical and chemical parameters of a matrix slightly depend on the chemical composition of the gel-formation media, in particular salts, which have basic value for many bacteria.

It should be noted that physiological concentration of some substances (acids, alkalis, amino acids, sugar, phosphates, carbonates, and Na salts) has no significant effect on physical characteristics of the carrier [16]. Physical characteristics of PVA gels depend not only on temperature but also on duration of the procedures and conditions of defrosting. The PVA itself performs the function of a cryoprotector.

A detailed study of the procedure of defrosting has shown that the high speed of defrosting (more than 10°C in a minute) can lead to the formation of colloidal solutions, while at slow defrosting (with a speed less 0.05°C in a minute), the elastic structures are formed. According to data of electronic microscopy, it is established that at slow thawing, the structural distinctions of the carrier is less.

High concentration of biomass of cells (more than 1 g/100 g) can render the destabilizing effect on carrier formation. In other concentrations, the structure of a matrix is optimal for cell stabilization. The diffusion into cryogel substrates, toxins, and various other chemicals occurs in macropores.

Physical characteristics of the created PVA gels are steady in the wide range of positive temperatures, up to 80°C, that allows high efficiency operation of the bioreactor in different temperature conditions.

The biotechnological processes, which are effectively proceeding with participation of the immobilized biological objects, were described. As examples, the formation of ethanol, acetate, hydrogen, transformation of sugars, amino acids, and some other important products of cellular metabolism has been studied [21, 22].

Thus, the physical and chemical properties, lack of toxicity, optical transparency, and formation of carrier happening at negative temperatures are extremely important viewpoints for receiving different types of PVA biosensors. At the same time, the limited number of works on the use of this carrier for immobilization of photobacteria or recombinant strains with lux-genes of bacterial luciferase system should be noted.

7. PVA immobilization of photobacteria

The immobilization of *Photobacterium phosphoreum* in PVA, polyurethane, polyacrylamide, Ca-alginate, and Ca-carboxy-methyl-cellulose is carried out [4, 14]. It was established [14] that photobacteria in PVA-gel possessed less stable, in comparison with other carriers, the duration of light emission—1 week.

The technology of immobilization in PVA-biofilm photobacteria *Vibrio fischeri*, and a gene engineering strain *Pseudomonas putida* with the lux-operon from *Photorhabdus luminescens* is realized for detection of phenolic toxins in industrial waste [24]. However, for increase in stability, light glycerin was introduced into the composition of test system; the procedure included activation of a luminescence with special chemicals.

The psychrophilic luminous bacteria *P. phosphoreum* were immobilized in PVA-cryogel. The technological procedures of immobilization allowed to preserve with preserve nearly 100% of level of emission activity of cells without protectors. The activation processes of luminescence are developed and optimized [13, 25, 26].

With the use of photobacteria as biosensor of toxins, special requirements are imposed to the duration and stability of a luminescence of a test system [13, 17]. These parameters, first of all, depend on natural emission properties of the chosen strain and are controlled by technology of an immobilization, the procedure of storage, and use of biosensors [3, 7, 20].

Specific activity at different strains of luminous bacteria is from 10^2 to 10^5 quanta/s. Light intensity and duration of a luminescent cycle are defined by a specific adaptable metabolism of bacteria. The luminescent cycle of deep culture *Vibrio harveyi* short was completely finished in a logarithmic growth phase (12–14 h), *Vibrio fischeri* has a cycle about 20–24 h, while *Photobacterium phosphoreum* has a stable luminescence at 1–2 weeks. The psychrophilic strains of photobacteria exhibit the highest specific activity and the most prolonged bioluminescence in the growth culture. The luminescent cycle of *Photobacterium phosphoreum* at 15–20°C can exceed 100 h [27, 28].

It is shown that the immobilization in PVA cryogel in all species of bacteria is capable to lead to essential prolongation of a luminescence [26].

It is obvious that the composition of gel-formation media should influence the cells' stability in the carrier and metabolic activity first of all, on stages of gel formation. For a choice of the optimal media composition and conditions, different mixtures have been used for gel formation, and the kinetics of light emission after "freezing/thawing" procedure was investigated. The best results have been received with media using for the liquid cultivation of photobacteria [29]. The emission activity of granules after a freezing/thawing procedure (24 h from the beginning the heating stage) has shown that in this case the immobilized bacteria were kept practically at 100% luminescent activity. The most essential recession of luminescence (in ~10^3 times) was observed when the gels' formation process comes with only 3% NaCl.

The immobilization procedure of photobacteria in PVA cryogel has been published in [26].

Bacterial strains	Immobilized cells		Free cells	
	Total light emission (Q), photons per cell	Time-course of luminescence, days	Total light emission (Q), photons per cell	Time-course of luminescence, days
P. phosphoreum (str. NZ-11D)	1–5 × 10^7	14	5 × 10^6	7
P. phosphoreum (str. №331 KM MSU)	1–5 × 10^9	42	10^3	21
V. harveyi (str. B392 MAV)	1–5 × 10^6	7	5 × 10^5	1
V. fischeri (str. №6 KM MSU)	1–5 × 10^6	10	10^5	3

Table 1.
The emission activity and duration of the free and PVA-immobilized different strains of photobacteria.

The developed technology has allowed the survival of luminous bacteria in the carrier, practically at 100%, without the introduction of cryoprotecting agents and procedures of a light induction. All strains of PVA-immobilized bacteria possess not only high level of initial emission but also raised stability of luminescent activity in comparison with the free cells (**Table 1**).

As the main results from the data presented, the integral light output (Q) is observed on preparations, which formed using the complex growth-media (GM) for the liquid cultivations. Free cells in the same conditions possess essentially smaller (more than 2 order), in comparison with immobilized cells, integral activity.

Specific bioluminescent activity was restored to level of activity of free cells (~10^5 quanta/s per cell). At storage, at −80°C, bioluminescent activity remains without changes for 2 years. The detected level of light emission is at 4°C for over 1 month and at 20°C for 3 days.

Thus, the composition of the gel formation media, not incubation mixture, makes the basic impact on intensity and duration of bioluminescence of immobilized preparations. Concentration of the carrier (5, 7, and 10%) practically does not change emission and kinetics parameters.

8. The kinetics of light emission free and immobilized bacteria *P. phosphoreum*

In **Figure 4**, the time dependence of specific luminescent activity during incubation at 4°C free and immobilized bacteria is presented. The immobilized cells (Curve 1) have more prolonged luminescence, than free bacteria (Curve 2). It is established that luminescence attenuation by immobilized bacteria is not a consequence of destruction of cells, or exhaustion of endogenic substrates, and reflects decrease in reduction potential of a cell owing to shift medium pH with products of a metabolism.

It was established that luminescence attenuation by immobilized bacteria is not a consequence of cell destruction, or exhaustion of endogenic substrates, and reflects decrease in reduction potential of a cell owing to shift medium pH with products of a metabolism.

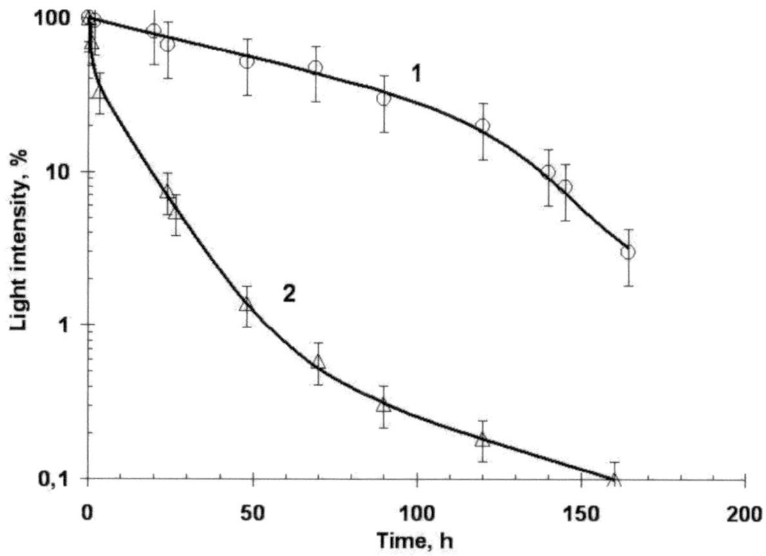

Figure 4.
Light output and bioluminescence decline from the PVA-immobilized and free cells of photobacteria following storage in 0.1 M Na-phosphate buffer +2% NaCl, pH 7.6, 4°C.

9. Photo-biomonitoring

The optimal conditions of reception and storage-immobilized cells formed a basis of use of luminescent granules as biosensor controls. The standard procedure of the analysis with use of free cells provides incubation with toxicants for 5, 15, and 30 min [29]. The analysis of inhibition kinetics from free and immobilized cells has shown similar time profiles, testifying to the absence of serious diffusion restrictions of a gel material and the form of granules for all types of molecules. Supervision logically follows from structural characteristics of the matrix having macropores. As essential distinctions in inhibition, kinetics is not revealed, and the time parameters postulated to free bacteria are chosen for the toxicity analysis with PVA-immobilized cells.

In **Table 2**, the granule luminescence inhibition by various classes of toxins, heavy metals, phenolic derivatives, and pesticides, is presented.

Before assay procedure granules were incubated 10-min with environment temperature (22°C), and the next 5–15 min with toxins. It is established that the threshold sensitivity of free and immobilized cells to the chosen quenchers of luminescence is approximately identical and as a whole corresponds to literary data.

Physical and geometrical parameters of photo-PVAG biosensors for continuous biomonitoring ecotoxicants with the minimum restrictions for diffusion of toxins are also optimized. In the technology of continuous monitoring of toxicants, the optimized conditions of the discrete analysis are used. It is established that in a channel mode, light activity remains about a week at solution temperature 10–12°C.

The kinetic profile of light inhibition depends on the concentration of toxicant. The luminescence time response of the PVA-immobilized cells in continuous flow-through monitoring to $CuSO_4$ is shown in **Figure 5**.

The kinetics is reflected by convertibility effect, although the level of luminescence after washing away toxicants can be different from the initial. The relaxation time increase with increasing the concentration of toxins. However, even at concentration exceeding EC_{50} 10 times at the chosen flow-rate, remains not less than 10% of

Toxicant	The range of toxin concentrations (mg/l) determined by bioluminescence response of PVA immobilized cells at different times of incubation	
	5 min	15 min
Cu^{2+}	5–40	1–8
Zn^{2+}	10–60	0.5–4
Hg^{2+}	0.1–0.6	0.05–0.10
2,4-Dinitrophenol	5–40	2–20
Pentachlorophenol	0.2–2.0	0.05–0.4
2,4-Dichlorophenol	0.5–3.0	0.5–10
2,4-Dimethylphenol	0.5–8.0	0.5–8.0
2,4-Dichlorophenoxyacetic acid	1.0–10.0	0.5–10.0
2,4,5-Thrichlorophenoxyacetic acid	0.5–4.0	1.0–8.0

Granules of different geometrical sizes (from 1 to 3 mm^3) were used.

Table 2.
The sensitivity of the PVA-immobilized cells P. phosphoreum to heavy metals, chlorinated phenolic derivatives, and pesticides (100 μl probe, 1 ml 3% NaCl with one PVA granule).

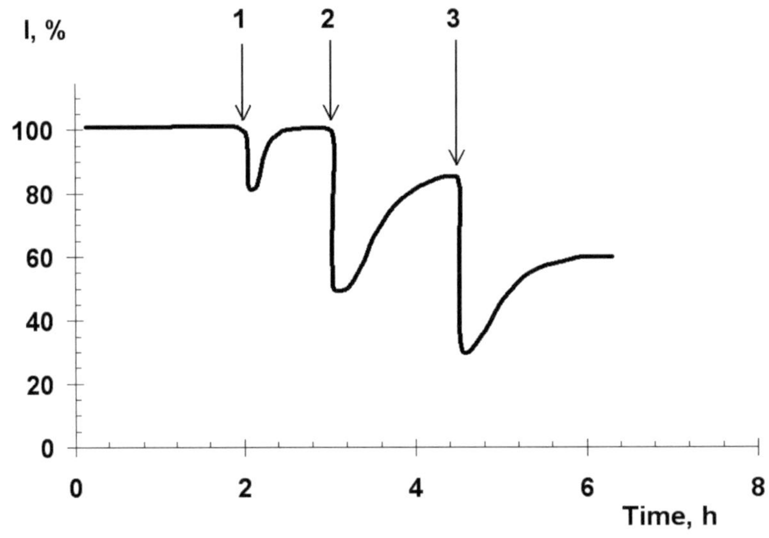

Figure 5.
The kinetic response of PVA-immobilized P. phosphoreum str. №331 KM MSU, in continuous flow-through monitoring system to $CuSO_4$, volume injected with 100 μl. Initial concentration: 1–0.5, 2–1, and 3–10 mM.

luminescent activity. Fast restoration of activity allows using one immobilized preparation for repeated probe detection (flow rate—0.5 ml/min, resistance time injected probe—30 s, reactor volume—1.5 ml, and 1 granule 15°C).

A continuous water toxicity system with same parameters for biomonitoring, using Ca,Sr-alginate gel immobilized *P. phosphoreum*, was also studied [16, 30].

Higher survival of cells with preservation of specific activity of light emission, at level of the free cells, presented to the given work, reflect advantages cryogenic immobilization photobacteria in PVA carrier and application PVA based biosensors for detection of toxicants both in discrete and continuous toxicity monitoring systems.

Author details

Anvar D. Ismailov[1]*, Leyla E. Aleskerova[1], Kristina A. Alenina[1]
and Elena N. Efremenko[2]

1 Biology Faculty MSU, Moscow, Russia

2 Chemical Faculty MSU, Moscow, Russia

*Address all correspondence to: anvaris@list.ru

References

[1] Durand MJ, Hua A, Jouanneau S, Cregut M, Thound G. Mint: Detection of metal and organometallic compounds with bioluminescent bacterial bioassays. Advances in Biochemical Engineering/ Biotechnology. 2016;**154**:77-99

[2] Ismailov AD, Aleskerova LE. Mint: Photobiosensors containing luminescent bacteria. Biochemistry (Moscow). 2015;**80**(6):733-744

[3] Lee JH, Mitchell RJ, Kim BC, Cullen DC, Gu MB. Mint: A cell array biosensor for environmental toxicity analysis. Biosensors & Bioelectronics. 2005;**21**(3):500-507

[4] Makiguchi N, Arita M, Asai Y. Mint: Immobilization of a luminous bacterium and light intensity of luminous material. Journal of Fermentation Technology. 1980;**58**(1):17-21

[5] Lei Y, Chen W, Mulchandani A. Mint: Microbial biosensors. Analytica Chimica Acta. 2006;**568**:200-210

[6] Brodelius P, Vandamme EJ. Mint: Immobilized cells systems. In: Rehm HJ, Reed G, editors. Biotechnology. A Comprehensive Treatise in 8 Volumes. Vol. 7a. Weinheim: VCH Verlag. pp. 405-464

[7] Park KS, Baumstark-Khan C, Rettberg P, Horneck G, Rabbow E, Gu MBM. Immobilization as a technical possibility for long-term storage of bacterial biosensors. Radiation and Environmental Biophysics. 2005;**44**:69-71

[8] Sinitsin AP, Raynina EI, Lozinskiy VI, Spasov SD. Immobilized Microbial Cells. [in Russian]. Moscow: Moscow State University Publishers; 1994. p. 288

[9] Heitzer A, Malachowsky K, Thonnard JE, Bienkowski PR, White DC, Sayler GS. Mint: Optical biosensor for environmental on-line monitoring of naphthalene and salicylate bioavailability with an immobilized bioluminescent catabolic reporter bacterium. Applied and Environmental Microbiology. 1994;**60**(5):1487-1494

[10] Ji-qiu Y, Li X-z, Chi Z, Yi-hua Z. Mint: Luminescent bacterial sensors made from immobilized films of *Photobacterium phosphoreum*. Chemical Research in Chinese Universities. 2005;**21**(1):44-47

[11] Kohler S, Belkin S, Schmid RD. Mint: Reporter gene bioassays in environmental analysis. Fresenius Journal of Analytical Chemistry. 2000;**366**(6-7):769-779

[12] Polyak B, Bassis E, Novodvorets A, Belkin S, Marks RS. Mint: Bioluminescent whole cell optical fiber sensor to genotoxicants: System optimization. Sensors and Actuators B: Chemical. 2001;**74**(1-3):18-26

[13] Arnold MA. Mint: Fiber-optic biosensors. Journal of Biotechnology. 1990;**15**:219-228

[14] Makiguchi N, Arita M, Asai Y. Mint: Optimum cultural conditions for strong light production by *Photobacterium phosphoreum*. Journal of General and Applied Microbiology. 1980;**26**:75-83

[15] Makiguchi N, Arita M, Asai Y. Mint: Optimal conditions for frozen storage of immobilized luminous bacteria. Journal of Fermentation Technology. 1980;**58**(4):333-337

[16] Chun U-H, Simonov N, Chen Y, Britzb ML. Mint: Continuous pollution monitoring using *Photobacterium phosphoreum*. Resources, Conservation and Recycling. 1996;**18**:25-40

[17] Kim SK, Lee BS, Lee JG, Seo HJ, Kim EK. Mint: Continuous water

toxicity monitoring using immobilized *Photobacterium phosphoreum*. Biotechnology and Bioprocess Engineering. 2003;**8**:147-150

[18] Cline TW. Mint: Isolation and characterization of luminescence system mutants in bacteria. Methods in Enzymology. 1978;**57**:166-171

[19] Kim BC, Gu MBM. A multi-channel continuous water toxicity monitoring system: Its evalution and application to water discharged from a pawer plant. Environmental Monitoring and Assessment. 2005;**109**:123-133

[20] Lee JH, Gu MBM. An integrated mini biosensor system for continuous water toxicity monitoring. Biosensors & Bioelectronics. 2005;**20**:1744-1749

[21] Lozinsky VI, Plieva FM. Mint: Poly(vinyl-alcohol) cryogels employed as matrices for cell immobilization. Overview of recent research and developments. Enzyme and Microbial Technology. 1998;**23**:227-242

[22] Lozinsky VI, Zubov AL, Kulakova VK, Vainerman ES, Rogozhin SV. Mint: Application of poly(vinyl alcohol) cryogels in biotechnology. II. Variation of theological properties of gel matrix as a result of yeast cells entrapment. Biotekhnologiya. 1990;**5**:32-35

[23] Varfolomeev SD, Rainina EI, Lozinsky VI, Kalyuzhnyi SB, Sinitsyn AP, Makhlis TA, et al. Application of poly(vinyl alcohol) cryogel for immobilization of mesophilic and thermophilic microorganisms. In: de Bont JAM, Visser J, Mattiasson B, Tramper J, editors. Physiology of Immobilized Cells. Wageningen: Elsevier; 1989. pp. 325-330

[24] Philp JC, Balmand S, Hajto E, Bailey MJ, Wiles S, Whiteley AS, et al. Mint: Whole cell immobilized biosensors for toxicity assessment of a wastewater treatment plant treating phenolics-containing waste. Analytica Chimica Acta. 2003;**487**:61-74

[25] Alenina KA, Aleskerova LE, Kascheyeva PB, Ismailov AD. Mint: The poly(vinyl alcohol)-immobilized photobacteria for toxicology monitoring. Engineering. 2012;**4**:118-119

[26] Efremenko EN, Senko OV, Kuts VV, Alenina KA, Kholstov AV, Ismailov AD. The luminescent biocatalyst for detecting toxicants. Patent RU. 2010:2394910

[27] Kuts VV, Ismailov AD. Mint: Physiological and emission characteristics of the luminescent bacterium *Photobacterium phosphoreum* from the White Sea. Microbiology. 2009;**78**(5):554-558

[28] Nealson KH, Hastings JW. Mint: Bacterial bioluminescence: Its control and ecological significance. Microbiology Review. 1979;**43**(4):406-518

[29] Bulich AA. Use of luminescent bacteria for determining toxicity in aquatic environments. In: Markings LL, Kimerle RA, editors. Aquatic Toxicology. Philadelphia: American Society for Testing and Materials; 1979. p. 98

[30] Cho J, Park K, Ihm H, Park J, Kim S, Kang I, et al. Mint: A novel continuous toxicity test system using a luminously modified freshwater bacterium. Biosensors & Bioelectronics. 2004;**20**:338-344

Section 2

Bioluminescent Organisms

Chapter 4

Ecological and Histological Notes on the Luminous Springtail, *Lobella* sp. (Collembola: Neanuridae), Discovered in Tokyo, Japan

Tadasu Sano, Yukimasa Kobayashi, Ikuko Sakai, Katsunori Ogoh and Hirobumi Suzuki

Abstract

Some species of springtail (Collembola) are luminous, but it is not known whether light emitted by springtail is due to self-luminescence, feeding on luminous fungi, or accidental infection by luminous bacteria. To address this question, we characterized the luminescence of a luminous springtail, *Lobella* sp. (family Neanuridae) discovered in Tokyo, Japan. The emitted light was yellowish-green (540 nm) and was found to originate from tubercles on the thorax (segments II and III) and abdomen (segments I–VI) using a low-light imaging system. The luminescence persisted for several seconds but showed occasional oscillations in a laboratory environment. We also observed fat bodies containing eosin-positive granules under the integument of the tubercles in the tergum by hematoxylin and eosin (HE) staining that were not present in a nonluminous springtail (*Vitronura* sp.). The fat bodies in *Lobella* sp. are presumably photocytes analogous to the firefly lantern, and the eosin-positive granules are the likely source of bioluminescence, which implies that springtails are self-luminescent.

Keywords: luminous springtail, *Lobella*, light organ, fat body, histology

1. Introduction

Some species of springtail (Collembola) are luminous. *Lipura noctiluca* [1], *Anurida* sp. [1], and *Anurida granaria* [2] of the family Neanuridae, *Anurophorus fimetareus* [1] of the family Isotomidae, and *Onychiurus armatus* [3] of the family Onychiuridae exhibit luminescence throughout the body and glow continuously, while *Neanura muscorum* [1] and *Neanura quadrioculata* [1] of the family Neanuridae emit flashes of light upon stimulation. The color of emitted light is diverse from bluish-green to greenish-yellow in each species. However, it is not known whether light emitted by springtail is due to self-luminescence, feeding on luminous fungi, or accidental infection by luminous bacteria. Springtails typically live in the soil of damp wooded areas that contains the mycelia of luminous fungi. The midgut of the luminous springtail species *Achorutes muscorum* (family Neanuridae) was found to be full of luminous mycelia, but species from other localities were nonluminous [3], suggesting that the light originates from fungi consumed by springtails. However, this is unlikely for species that glow only when stimulated. On the other hand, *Onychiurus armatus* exhibited luminescence even after rearing for 6 months on sterilized agar, providing support for the self-luminescence hypothesis [3]. However, to date there are no detailed descriptions of the histology of light organ and chemistry of luminescence in luminous springtails [3–5].

One of authors (Sano) discovered a luminous springtail, *Lobella* sp., belonging to the family Neanuridae in Tokyo, Japan [6]. This species emits light not from the whole body but from spots on the abdominal segments, which has never been previously reported in springtail [3–5]. This makes *Lobella* sp. suitable for histological studies of a potential light organ. In this work we describe the habitat of *Lobella* sp., the nature and origin of the luminescence, and histological findings.

2. Materials and methods

2.1 Springtails

Lobella sp., which is closely related to *Lobella sauteri*, was collected from two sites: (1) Yokosawa (Akiruno-shi, Tokyo, Japan) and (2) Mitake (Oume-shi, Tokyo, Japan). A nonluminous species, *Vitronura* sp. (family Neanuridae) collected from Takiyama (Hachioji-shi, Tokyo, Japan), served as a control.

2.2 Luminescence imaging

The luminescence of the specimen was captured on Fujicolor 1600 film (Fujifilm, Tokyo, Japan) with a Nikon FM camera (Nikon, Tokyo, Japan) with Nicol 50-mm (F1.4) and 2× teleconverter lenses. Luminescence images were also acquired with a luminescence microscope using a short focal length lens system [7, 8] equipped with an ImagEM electron-multiplying charge-coupled device (CCD) camera (C9100-13; Hamamatsu Photonics, Hamamatsu, Japan). The total magnification was reduced from 4× (UPLSAPO4× objective lens; Olympus, Tokyo, Japan) to 0.8× using the short focal length imaging lens (f = 36 mm, NA = 0.2) in order to capture low intensity light.

Video recording of the luminous specimen was performed using a C2400 high-resolution SIT video camera system (Hamamatsu Photonics) equipped with a Zuiko auto-zoom 35–105 mm (F3.5–4.5) lens (Olympus). The video was converted to an avi format file, and time-lapse image analysis was performed using TiLIA software [9] in order to determine the time course of luminescence intensity in a region of interest.

2.3 Spectroscopy

The luminescence spectrum of the specimen was determined using a U-2900 spectrometer (Hitachi High-Technologies, Tokyo, Japan) under the following conditions: transmittance mode without incident light; scan range, 450–650 nm in 5-nm steps; scan speed, 240 nm/min; and response time, 2 s.

2.4 Histology

The whole body of each specimen was fixed in alcohol Bouin's solution and stored at room temperature. The sample was dehydrated in a graded series of ethanol, embedded in paraffin, and sectioned at a thickness of 5 μm with a microtome (Microm HM310; Leica Biosystems, Nussloch, Germany). The sections were stained with Delafield's hematoxylin and eosin (HE). A BX53 microscope (Olympus) with a DP74 color CCD camera (Olympus) was used for light microscopy observation.

3. Results

3.1 Habitat description

Lobella sp. was collected at Yokosawa, which is a hill region (200–300 m in height) surrounded by valleys that is a typical secondary forest used for agriculture (**Figure 1A**). In one area, shiitake mushrooms (*Lentinula edodes*) are cultured on logs on the ground surface of the forest. After the mushrooms are harvested, the logs proceed to decay (**Figure 1B**). *Lobella* sp. was found under or in the decaying logs (**Figure 2A**) from April to the end of October in 1997–2004. Luminous mushrooms of an unidentified species were also present in this area. *Lobella* sp. was also collected from a mountain pass beside Mount Mitake (900 m in height); the specimens were found under litter, particularly from *Magnolia obovata* (family Magnoliaceae) (**Figure 2B**). Although the luminous springtail, *Lobella* sp., inhabits in such different habitats, this study does not consider taxonomically whether they

Figure 1.
Habitat of the luminous springtail (Lobella sp.). (A) Overview of the habitat at the Yokosawa site (Akiruno, Tokyo, Japan). (B) Shiitake mushroom (Lentinula edodes) is cultured on logs, which proceed to decay on the ground after the mushrooms are harvested. Springtails are found under or on decaying logs.

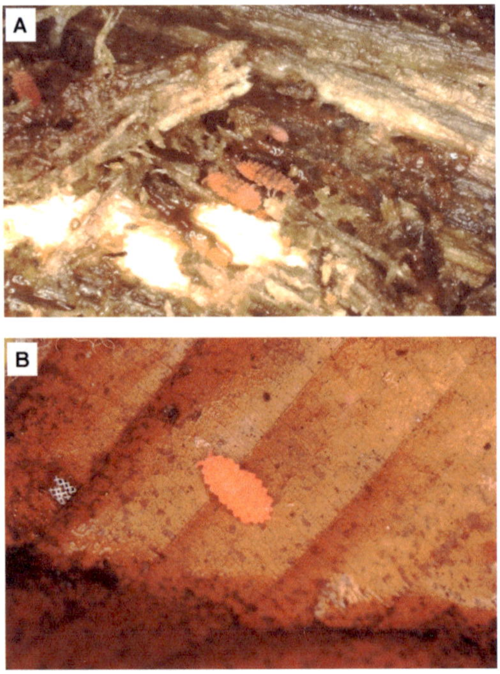

Figure 2.
(A and B) luminous springtail (Lobella sp.) found in decayed logs at the Yokosawa site (A) and under Magnolia obovata *(family Magnoliaceae) litter at the Mitake site (B).*

Figure 3.
Habitat of nonluminous springtail (Vitronura sp.) collected at the Takiyama site (Hachioji, Tokyo, Japan). (A) Cattle barn. (B) Vitronura sp. inhabits soil-containing cattle dung.

are the same or different species. The nonluminous springtail *Vitronura* sp. was collected at Takiyama (**Figure 3A**), a cattle barn, from soil-containing cattle dung (**Figure 3B**).

3.2 Luminescence

In a field observation at the Yokosawa site, we observed flashes of light on the ground when lifting the shiitake logs. However, we were unable to discern the color of the light by the naked eye since it was too weak. **Figure 4** shows *Lobella* sp. and its luminescence. Several tubercles were present on the surface of the thorax and abdomen (**Figure 4A**), and the luminescence was strongest at the posterior margin of the abdomen (**Figure 4B**). The luminescence was elicited for several seconds by mild mechanical stimulation (e.g., by shaking the container or blowing on the specimen), although there were long intervals between bursts of light. The luminescence spectrum showed a maximum peak wavelength of 540 nm (**Figure 5**), which is visible as a yellowish-green color. Mechanical stimulation also induced the secretion of mucus from the body surface; at such instances, no luminescence was observed either from the body or the mucus.

To identify the luminescent region of the body, luminescence image of the specimen was captured by microscopy (**Figure 6**). The luminous spots corresponded to tubercles on the thorax (segments II and II) and abdomen (segments I through VI). Time-lapse image analysis showed that the luminescence persisted for several seconds (less than half a minute) in the laboratory environment with occasional oscillations with a 3.0-s flash interval, 2.1-s pulse duration, 0.9-s inter-pulse duration, and 3.0-s oscillatory peak interval (determined from the average of four peaks in **Figure 7**).

3.3 Histology

HE staining of a cross section of the second abdominal segment of *Lobella* sp. from the Yokosawa site revealed that the midgut cavity occupied most of the body and the tubercles had a pointy seta within a socket (**Figure 8**). The dorsal vessel (heart), ventral nerve cord, hemocytes, and muscle tissues were also identified. Under the integument of the tubercle in the tergum, fat bodies occupy a large space between the integument and midgut and comprise large trophocytes whose cellular boundary is obscure. Each trophocyte contains many eosin-positive granules (EPG) in its peripheral cytoplasm. Such granules were not present in fat bodies under the

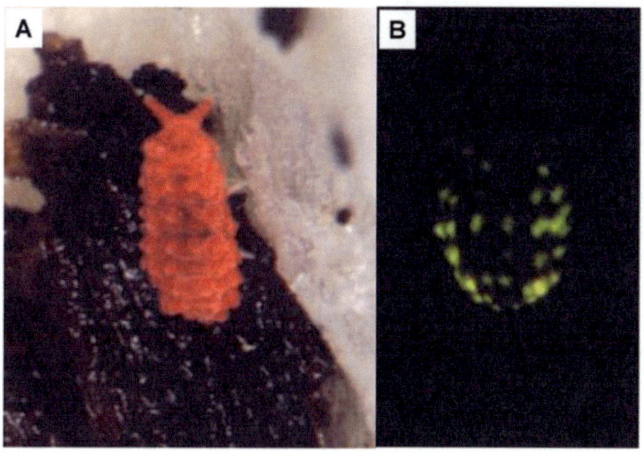

Figure 4.
Photographs of Lobella sp. (A) Bright-field image of Lobella sp. collected at the Yokosawa site. Tubercles are present on the body surface of the thorax and abdomen. (B) Bioluminescence images of Lobella sp. collected at the Mitake site. Luminescent spots are present on the body surface; a higher signal intensity is observed at posterior margin of the abdomen.

integument of the sternum or pleuron. The posterior fat bodies were larger than those in the anterior region and were surrounded with hemolymph (**Figure 9**). No eosin-positive granules were observed in fat bodies of *Vitronura* sp. (**Figure 10**).

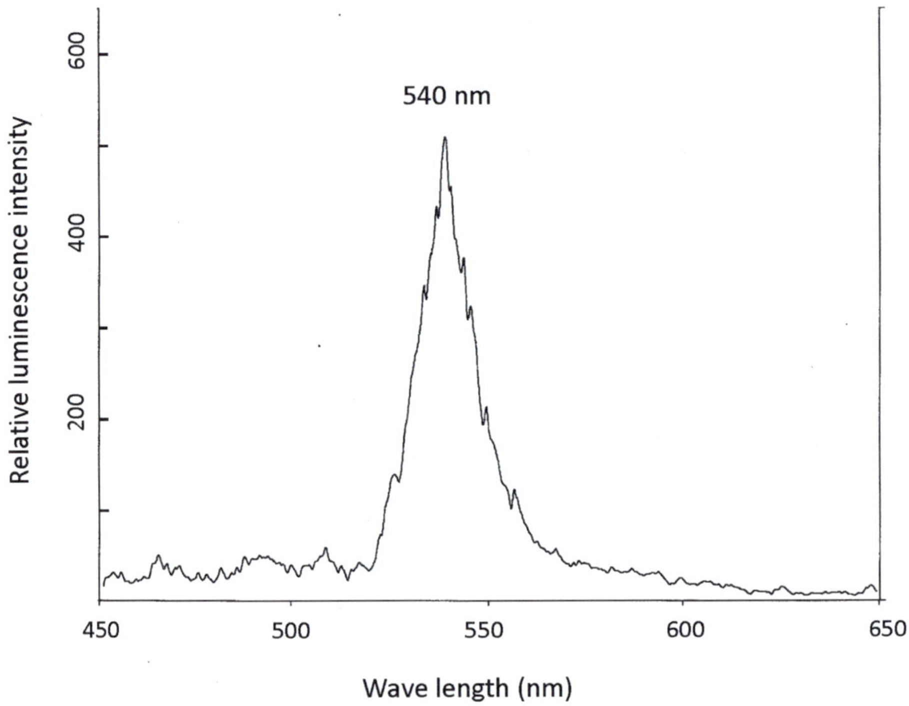

Figure 5.
Bioluminescence spectra of Lobella sp.; the maximum peak wavelength is 540 nm.

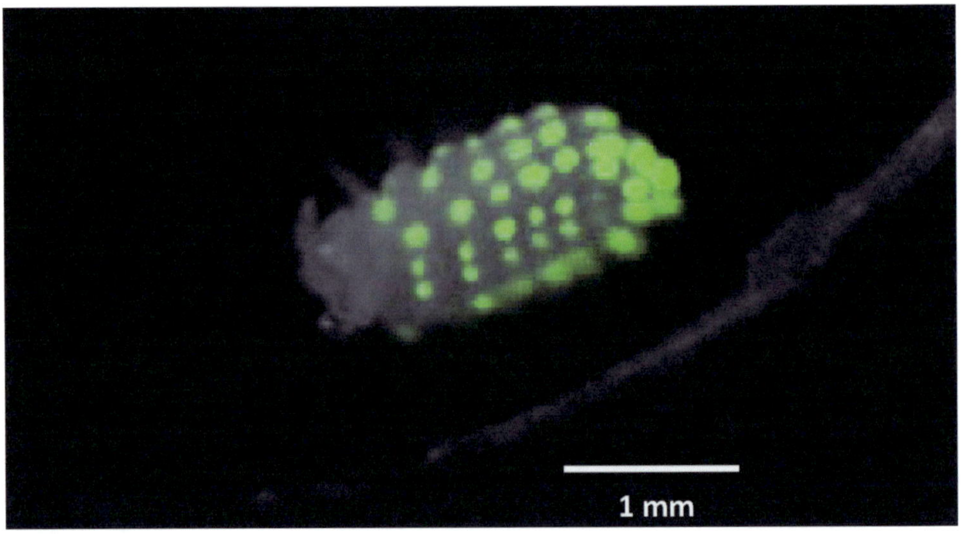

Figure 6.
Merged bioluminescence and bright-field images (pseudocolored green) captured with the luminescence microscope. The luminous spots correspond to tubercles on the thorax (segments II and III) and abdomen (segments I–VI). Scale bar: 1 mm.

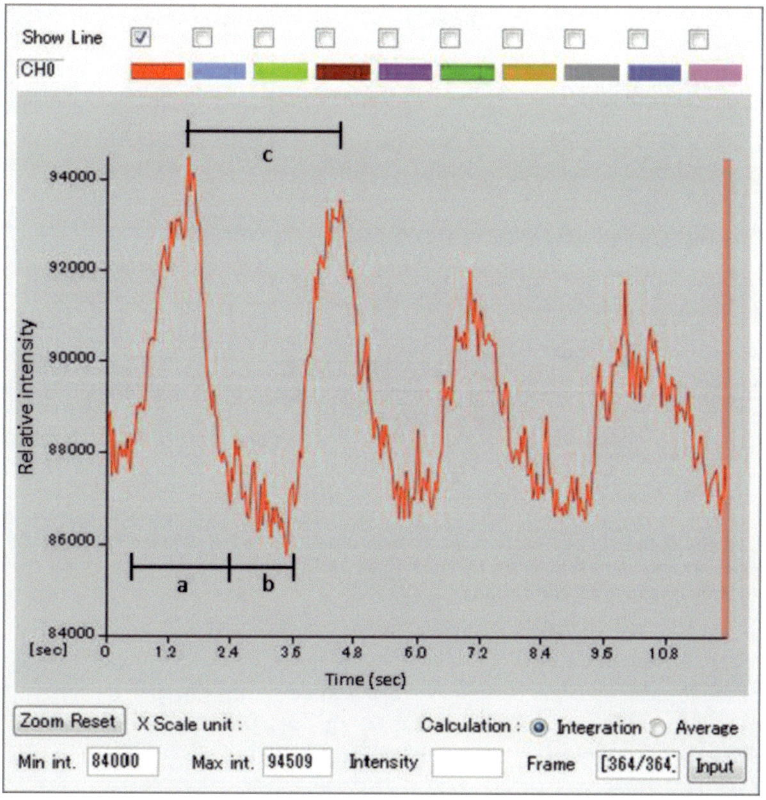

Figure 7.
Time course of luminescence captured by time-lapse imaging. The X and Y axes show time(s) and relative luminescence intensity, respectively. Occasional oscillations were observed, and the flash interval (a + b) is 3.0 s with 2.1 pulse duration (a) and 0.9 s inter-pulse duration (b), and oscillatory peak interval (c) is 3.0 s.

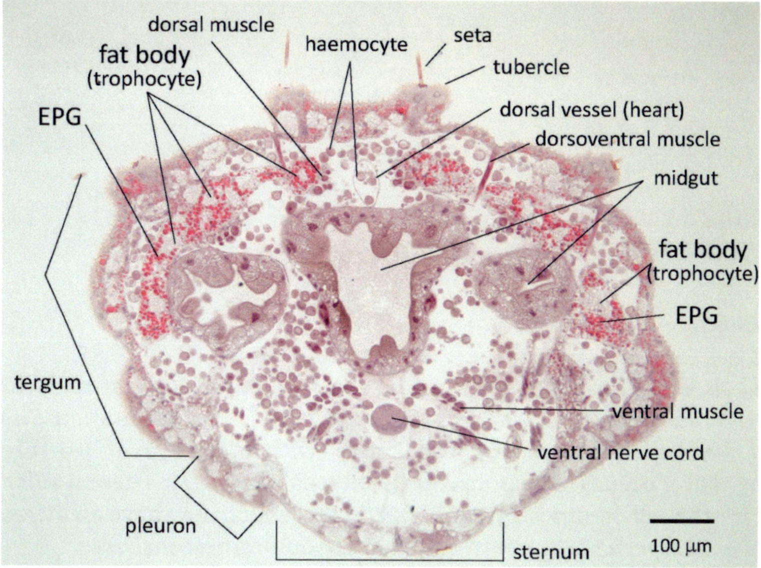

Figure 8.
Cross section of the second abdominal segment visualized by HE staining of Lobella sp. from the Yokosawa site. Fat bodies containing eosin-positive granules (EPG) are present under the integument of the tubercles in the tergum. Scale bar, 100 μm.

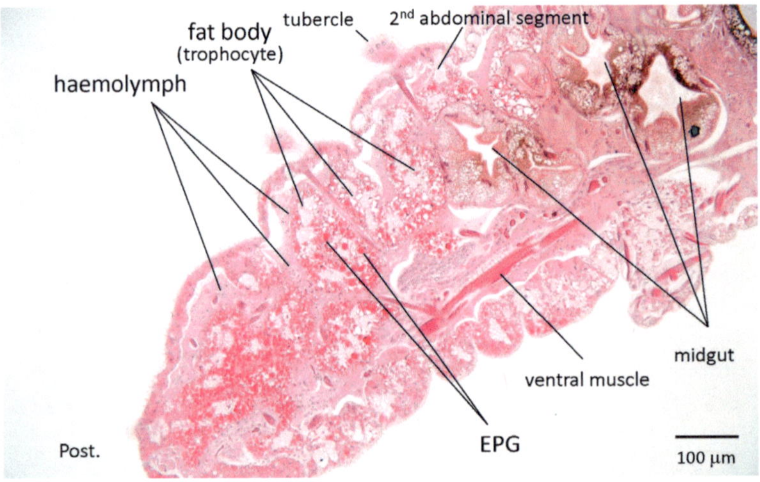

Figure 9.
Longitudinal section of Lobella sp. from the Yokosawa site visualized by HE staining. The fat bodies in the posterior region are larger than those in the anterior region and contain eosin-positive granules (EPG) surrounded by hemolymph. Scale bar, 100 μm.

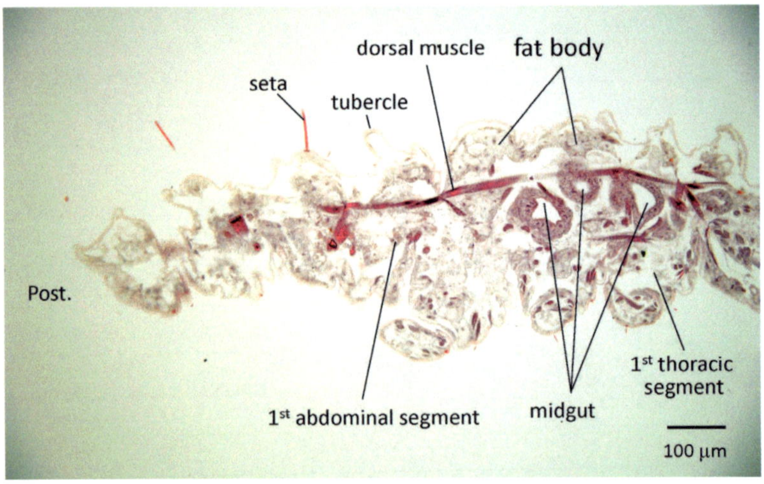

Figure 10.
Longitudinal section of Vitronura sp. from the Takiyama site visualized by HE staining. No eosin-positive granules were observed in the fat body. Scale bar, 100 μm.

4. Discussion

Lobella sp. emits flashes of light lasting several seconds upon stimulation. This is similar to *Neanura quadrioculata* and *N. muscorum* [1]. Furthermore, the luminescence of *Lobella* sp. was occasional oscillatory (with an interval of 3 s) (**Figure 7**), suggesting that it occurs in response to an external factor [1]. These results provide evidence for the self-luminescence of *Lobella* sp., which may have an intracellular origin since this species secretes mucus that is nonluminescent.

The light emitted by *Lobella* sp. was associated with tubercles on the abdomen (**Figure 6**), which may be light organs analogous to those of the firefly. A histological examination revealed fat bodies containing eosin-positive granules under the integument of the tubercles in the tergum (**Figures 8** and **9**), but we did not recognize any specific structures such as a photogenic or reflector layer of firefly

DOI: http://dx.doi.org/10.5772/intechopen.88321

that could function as a light organ. Fat bodies were also present in the nonluminous springtail *Vitronura* sp., but these did not contain eosin-positive granules (**Figure 10**). Photocytes of the light organ in larval and adult fireflies are derived from fat bodies [10–13], and luciferase is located in the cytoplasm of photocytes in the photogenic layer [14]. Furthermore in a larval firefly, fat bodies are a precursor of photocytes that were observed to have some luminosity by ex vivo low-light microscopy [15]. We speculate that the eosin-positive granules in the fat bodies of *Lobella* sp. are related to the bioluminescence of this species in the same manner as peroxisomes of firefly photocytes [13]. Ex vivo low-light imaging of fat bodies could confirm this possibility. On the other hand, an anti-firefly luciferase antibody did not react with *Lobella* sp. tissue (data not shown), indicating that the bioluminescence of springtail is distinct from the luciferin-luciferase reaction in firefly.

There have been few studies on luminous springtails, and most of these have been review articles [2–5]. Some reasons for the lack of research are the difficulty of species identification by entomologists who do not study Collembola and the small size of specimens, which makes biochemical analyses challenging. The present findings provide a basis for future studies on the mechanisms as well as the evolution of bioluminescence in insects using a variety of experimental approaches—e.g., molecular modeling, genetic engineering, and omics technology with artificial intelligence processing—that are applicable to small-sized organisms and wild specimens [16–19]. Additionally, springtail displays a unique mating behavior where the male deposits spermatophores on the ground or close to a female and keeps other males away through jostling [20, 21]. It is possible that the bioluminescence of springtail plays an important role in mating, as is the case for firefly and other luminous organisms [22].

Acknowledgements

We thank the late Prof. Ryosaku Ito (Showa University, Japan) for identifying the luminous springtail used in this study and the late Dr. Masayasu Konishi (Koganei City, Tokyo, Japan) for enabling our communications with Professor Ito and for valuable discussion.

Conflict of interest

The authors declare no conflict of interest.

Author details

Tadasu Sano[1], Yukimasa Kobayashi[2], Ikuko Sakai[3], Katsunori Ogoh[3] and Hirobumi Suzuki[4]*

1 Sano Apiculture Garden, Tokyo, Japan

2 Tokyo Metropolitan University, Tokyo, Japan

3 Evaluation Technology Department 1, Olympus Corporation, Tokyo, Japan

4 Research and Innovation Department, Olympus Corporation, Tokyo, Japan

*Address all correspondence to: hirobumi2_suzuki@ot.olympus.co.jp

References

[1] Barber HS. Luminous Collembola. Proceedings of the Entomological Society of Washington. 1913;**15**:46-50

[2] Malakauskiene A. Reported and potential bioluminescent species in Lithuania. Biologija. 2018;**64**:181-190

[3] Harvey EN. Bioluminescence. New York: Academic Press; 1952. p. 649

[4] Haneda Y. Luminous Organisms. Tokyo: Kouseisha-Kouseikaku; 1985. p. 318. (in Japanese)

[5] Oba Y, Branham MA, Fukatsu T. The terrestrial bioluminescent animals of Japan. Zoological Science. 2011;**28**:771-789

[6] Kashiwabara S. Luminescence of springtail: An origin of firefly luminescence? SCIaS. 1997;**2**(15):10. (in Japanese)

[7] Ogoh K, Akiyoshi R, May-Maw-Thet, Sugiyama T, Dosaka S, Hatta-Ohashi Y, et al. Bioluminescence microscopy using a short focal-length imaging lens. Journal of Microscopy. 2014;**253**:191-197

[8] Suzuki H, May-Maw-Thet, Hatta-Ohashi Y, Akiyoshi R, Hayashi T. Bioluminescence microscopy: Design and applications. In: Thirumalai J, editor. Luminescence—An Outlook on the Phenomena and Their Applications. Rijeka, Croatia: Intech; 2016. pp. 333-349

[9] Konno J, Hatta-Ohashi Y, Akiyoshi R, Thancharoen A, Silalom S, Sakchoowong W, et al. TiLIA: A software package for image analysis of firefly flash patterns. Ecology and Evolution. 2015;**6**:3026-3031

[10] Williams FX. Photogenic organs and embryology of lampyrids. Journal of Morphology. 1916;**28**:145-207

[11] Hess WN. Origin and development of the light-organs of *Photurus pennsylvanica* de Geer. Journal of Morphology. 1922;**36**:245-277

[12] Buck JB. The anatomy and physiology of the light organ in fireflies. Annals of the New York Academy of Sciences. 1948;**49**:397-470

[13] Case JF, Strause LG. Neurally controlled luminescent systems. In: Herring PJ, editor. Bioluminescence in Action. New York: Academic Press; 1978. pp. 397-482

[14] Nur KS, Nurul WO, Norela S. The localization of luciferase in *Pteroptyx tener* (Coleoptera: Lampyridae) light organ. Serangga. 2016;**21**:51-59

[15] Tonolli PN, Okawachi FM, Abdalla FC, Viviani VR. Bioluminescent fat body of larval *Aspisoma lineatum* (Coleoptera: Lampyridae) firefly: Ontogenic precursor of lantern's photogenic tissue. Annals of the Entomological Society of America. 2011;**104**:761-767

[16] Fallon TR, Lower SE, Chang CH, Bessho-Uehara M, Martin GJ, Bewick AJ, et al. Firefly genomes illuminate parallel origins of bioluminescence in beetles. eLife. 2018;**7**:e36495

[17] Lower SE, Stanger-Hall K, Hall DW. Molecular variation across populations of a widespread north American firefly, *Photinus pyralis*, reveals that cording changes do not underlie flash color variation or associated visual sensitivity. BMC Evolutionary Biology. 2018;**18**:129

[18] Kotlobay AA, Sarkisyan KS, Mokrushina YA, Marcet-Houben M, Serebrovskaya EO, Markina NM, et al. Genetically encodable bioluminescent system from fungi. Proceedings of the National Academy of Sciences

of the United States of America. 2018;**115**:12728-12732

[19] Gregor C, Gwosch KC, Sahl SJ, Hell SW. Strongly enhanced bacterial bioluminescence with the ilux operon for single-cell imaging. Proceedings of the National Academy of Sciences of the United States of America. 2018;**115**:962-967

[20] Hopkin SP. Biology of the Springtails (Insecta: Collembola). Oxford: Oxford University Press; 1997. p. 330

[21] Schaller F. Indirect sperm transfer by soil arthropods. Annual Review of Entomology. 1971;**16**:407-446

[22] Lloyd JE. Bioluminescence and communication in insects. Annual Review of Entomology. 1983;**28**:131-160

Chapter 5

Effect of Camera Illumination on Flashing Behavior of *Pteroptyx malaccae* (Coleoptera: Lampyridae)

Anchana Thancharoen and Sirima Masoh

Abstract

Pteroptyx malaccae is a synchronous firefly that is important in firefly tourism in Thailand. Without well-managed tourism, the fireflies have faced to the problems of shooting camera flashes from tourists. Although the effect of artificial light was well understood, which causes negative impact to firefly courtship, there is no obvious information on the effect of the camera illumination. The experiment of testing four types of camera illumination was set up in laboratory using wild populations of *P. malaccae*. The flash patterns were recorded by videotaping and analyzed by using TiLIA software. The results showed that all kinds of camera illuminations affect flashing behavior of the fireflies. They prolonged flash interval by increasing pulse duration. The flashes from smartphone camera displayed the strongest effect; however, all flash types did not influence on the firefly life span, mating behavior and oviposition behavior of the fireflies.

Keywords: Lampyridae, synchronous firefly, firefly tourism, camera flashes, light pollution, TiLIA software, camera photography

1. Introduction

Recently, artificial night lighting occurs commonly in many urban areas and can be light pollution that influences negatively in many ways, i.e., waste the energy for the pollution production, causing sky glow, creating light trespass, and building glare [1] including causes of ecological effects on many organisms [2–6]. The night lighting changed unnaturally the innate behaviors of the organisms by reducing foraging behavior, predatory behavior, metabolism, growth rate, and reducing population numbers finally. Additionally, many species of insects were unavoidably trapped by phototaxis ability [4, 7, 8] including fireflies. Unfortunately, the fireflies' bioluminescent courtship signals were likely interfered [4].

The effect of artificial light on fireflies was studied in many firefly species. Under the light pollution, the abundance of fireflies displayed lower numbers than without lighting [4, 9]. Several firefly researchers have considerably examined how the light pollution impacts on the bioluminescent insects especially their courtship and reproduction. The flooding of light altered mate location behavior of *Lampyris noctiluca* males by avoiding seeking females in the lighted background areas where the female glows were less attractive [10]. It decreased the flashing activity of *Photuris versicolor* and mating success in *P. pyralis* [4]. The prolonged courtship time during dorsal mounting posture was significantly observed in *Sclerotia aquatilis* that probably results in failure of mating [11]. However, the sexual communication of some firefly species could happen in light-polluted area. The males of *A. ficta* shifted the flash signals to attract females by increasing light intensity and flash frequency in *A. ficta* [12]. Nevertheless, the mating success was not discussed. The variation of the impacts might depend on lighting types and firefly species.

Camera illumination or flash photography is another type of man-made light that is commonly used. Although the cameras do not produce long steady light, the blight flashlight probably influences animal behaviors. The effects of flash photography on organisms have been little studied. The flash testing on seahorse behaviors underwater was done and showed no significant long-term consequences [13]. The nocturnal insects such as fireflies have never been investigated.

Congregation of male *Pteroptyx* fireflies exhibited impressively synchronous courtship flashes to attract the females perching on mangrove trees. Many *Pteroptyx* habitats have been developed to be tourist attraction sites and improved local people's income in Thailand. Many consequences of firefly tourism have been considerably occurring in Thailand because of lacking knowledge of firefly ecotourism management [14]. Using flash photography of tourists might be adverse to firefly behavior, while the actual impacts have remained unknown.

In the present study, we examined the effects of camera illumination on flashing behavior of male and female *P. malaccae* under laboratory condition. Four treatments of commonly used flash photography were tested. We analyzed the fireflies' flashes by using computer software, time-lapse image analysis (TiLIA), to investigate actually the abnormal flashing behavior to expect the effect of the light on the real behavior of the fireflies in nature.

2. Materials and methods

2.1 Study species

2.1.1 Taxonomy

P. malaccae is one of more than 18 described *Pteroptyx* species distributed mostly in the region of Southeast Asia (from Boneo, Cambodia, Malaysia, Philippines, Sulawesi, and Thailand) through Australia. Their major characters for species identification are yellowish brown elytra with black deflexed elytral apices, pale brown abdominal ventrites, and dark brown head and antennae. Significantly, their males have metafemoral combs (MFC) on metathoracic legs, bipartite light organ in ventrite 7, and no lobes along the posterior margin of tergite 8 [15, 16]. From reference [17], they found morphological variation among *P. malaccae* specimens in different regions and classified them into four groups. In the group occurring in Thailand, the males have been sculpturing at the apices of the posterolateral projections (PLP) on ventrite 7. The females have similar color patterns as the males but have normal straight wing apices.

2.1.2 Ecology

P. malaccae inhabited in mangrove forest and backish water area. The larvae live on moist soil or mud and predate mangrove snails for food. At nighttime, the adults congregate on mangrove trees, or other vegetations grow nearby and flash synchronously to display their advertising flashes to their mates. *P. malaccae* is the most common synchronous species found in Thailand when compared with other three *Pteroptyx* species (*P. valida*, *P. tener*, and *P. asymmetria*). In many observations *P. malaccae* species in Thailand are usually found in the same colony with small numbers of *P. valida*. The identification between these two sympatric species can be done by both morphology and flash pattern (**Figure 1**). The flashes of *P. malaccae* are regularly shorter and faster (0.08–0.12 s of pulse duration with about 0.45 s interval), while *P. valida* flashes are brighter and longer (0.2–0.8 s of pulse duration with about 0.7–6 s interval) [18].

2.1.3 Firefly collection and maintenance

Two hundred adults of *P. malaccae* were collected from mangrove trees along Bangpakong riverbank, Chachoengsao province, after sunset time by using sweep nets. The females are difficult to observe because of having small light and low active behaviors; the proportion of the captured females was commonly lower than males. The sweeping at the low level of vegetation area of the mangrove trees where the females distribute preferably could capture the considerable numbers of them. The sexes of the fireflies were separated to maintain virgin status of adult females as possible. They were feeding on 10% honey solution to be healthy and have long life span.

Because of the congregation behavior of *Pteroptyx* fireflies on mangrove trees, four males and one female were randomly put in the same experimental box to observe mating communication between sexes and synchronous flashing behavior among males. The small group of males could display synchronous flashing behavior. Before experimental testing, each set of fireflies were randomly grouped in a 7.10 × 11.04 × 6.50 cm of transparent plastic box with a small moist cotton and allowed them to have an adaptation period for 15–30 min before starting the experiment.

2.2 Experimental design

Four treatments of different light sources of camera illumination were set up in laboratory: (1) a white flash from smartphone camera (SC), Samsung Galaxy Note 3; (2) a white flash from digital camera (DC), SLR Olympus TG 4; (3) red light for autofocus assist before a white flash from digital camera (RDC), Sony Exmor R; and (4) no flash (control). These treatments were selected from the representative flash characters found in firefly photography.

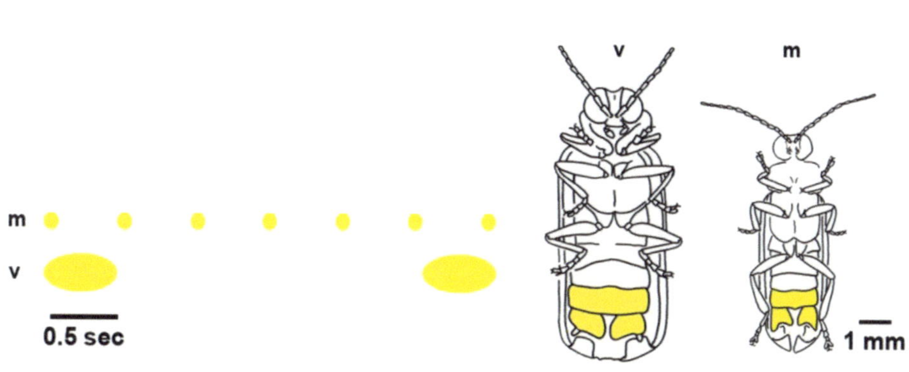

Figure 1.
Comparison of the sympatric firefly species, P. valida (v) and P. malaccae (m).

The apparatus of the experiment were designed as in **Figure 2**. The distance between the illumination point and the tested fireflies was 2 meters that is the possible distance in the real photography situation of tourists. The firefly flashing behavior was videographed using Sony Handycam™ digital camera recorder (HDR-SR11E) for 20 min in each replication (10 replications per treatment).

In the real cases of firefly photography by tourists, the fireflies would face with sequences of camera illumination; the tested fireflies were experimentally exposed to 3-time flashing to examine a consequence of sequential flashing. The flashes were shot after 5 min in the beginning of the experiment and were shot 3 times with 5-min intervals. The fireflies were continually observed and allowed to lay eggs after finishing the experiment. The adult lifespan and number of the eggs were recorded. The fireflies were tested within four nights after collection to avoid the errors from weakness and aged adults.

2.3 Flash analysis

The one of four males in each experimental set were randomly selected to analyze flash patterns. The video files were converted to Audio Video Interleave or .AVI format file to analyze by using time-lapse image analysis (TiLIA), a free software package for signal and flight pattern analyses of fireflies (available at Google Drive: https://drive.google.com/open?id=0B2o7FRVs2VohMmx2QzBVX3ZDeDA) [19]. Each video frame was converted to Tagged Image File Format (.TIFF) at each 0.03 sec. The light organs of all fireflies were defined as area of interest (ROIs) to measure the light intensities. The outputs (time and light intensities) were exported to Microsoft Excel® for flash parameter calculation applied from reference [20] (**Table 1**).

The flashes during courtship time or during synchronous flashing of the tested males were selected for flash parameter analysis by counting numbers of frames occurred in each flash parameter—pulse duration, interpulse duration, and flash interval—and then converted to second time unit.

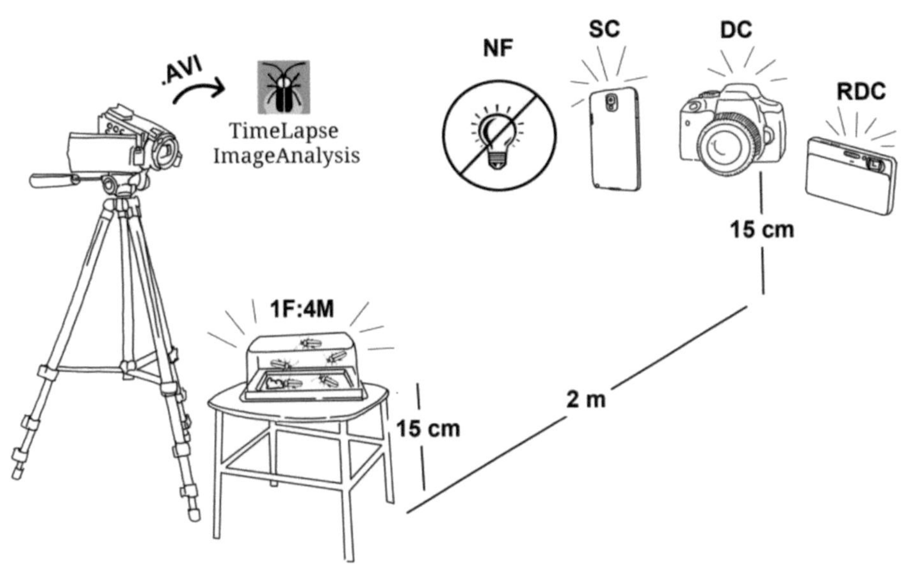

Figure 2.
Experimental apparatus.

Flash parameters	Explanation
Pulse duration	Duration between the beginning flash to the end of the flash (light period)
Interpulse duration	Duration between two pulse durations (dark period)
Flash interval	Summarizing of pulse and interpulse durations
Flash rate	Number of flash per time

Table 1.
Terminology of flash signals.

2.4 Statistical analysis

Thirty flashes of males in each treatment were statistically analyzed. The comparison of three types of flash parameters among camera illumination treatments and among different time sequences was statistically analyzed using one-way ANOVA and Tukey multiple comparison test. The flash rate of smartphone treatments was compared with control group by using independent t-test. A value of $p < 0.05$ was considered statistically significant. All statistical analyses were performed using SPSS program version 24.

3. Results

3.1 Flash parameters of *P. malaccae* males

A thousand and 33 flashes from 30 males in the control group were analyzed for flash parameters of *P. malaccae*. The *P. malaccae* males displayed variation of flash patterns. Their pulse duration could be in the range of 0.06–0.24 with 0.15–0.85 s of interpulse duration (**Table 2**). There was no correlation between pulse and interpulse duration (Pearson correlation coefficient $r = 0.033$, $P = 0.288$, $n = 1023$).

3.2 Effect on flash parameters

The flashing behavior of *P. malaccae* fireflies is behavioral plasticity; they could change their flash signals in response to the artificial light stimuli. From the results of flash parameter analysis, the artificial illumination from all camera flash types caused flash parameter plasticity by extending pulse duration significantly ($F = 68.461$, $df = 3$, $P = 0.000$), SC > DC > RDC > NF, and resulted in prolonging a long duration of interpulse found only in the case of smartphone camera flash ($F = 44.494$, $df = 3$, $P = 0.000$) (**Figure 3**). These resulted in longer flash interval of the smartphone treatment than the normal flash patterns. The tested fireflies in smartphone flash treatment displayed extremely slow flash rate when compared with the control group ($t = -6.346$, $df = 1682$, $P = 0.000$).

The results of the study showed that camera flashes affect temporarily the flashing behavior of *Pteroptyx* fireflies. After shooting all types of the camera flashes, most male fireflies displayed paused flashing for a period (approximately 3 sec) and then become flashing normally. The levels of sensitivity depended on flash types; smartphone camera has the highest effect (60%) followed by the digital camera with red light for focusing (30%) and digital camera (10%), respectively. Although the fireflies were shocked by the flashes suddenly, approximately 80–90% of male fireflies in all flash treatments could display courtship flashes and posted dorsal mounting during mating behavior later. Some of the males could reveal synchronous flashing behavior, DC (30%) > RDC (20%) > SC (10%).

3.3 Different time series of the flashes

After applying each types of flash photography treatment 3 times, all flash parameters in different time series were analyzed and compared in each flash type. The DC and RDC treatments displayed fluctuation of pulse duration but did not quite differently before shooting flash (**Figure 4a**). On the other hand, SC displayed an increasing trend of pulse duration ($F = 62.899$, $df = 3$, $P = 0.000$). The pulse duration after the last exposure of flash was the highest at 0.15 ± 0.00 s.

The flash series did not affect interpulse duration and flash interval of DC, although they could cause a small fluctuation in RDC. Alternatively, SC increased interpulse duration and flash interval after receiving the first flash ($F = 34.345$, $df = 3$, $P = 0.000$; **Figure 4b**) before flashing normally when received later flashes.

3.4 Female responses

The female flashes are not likely the answer response to the male flashes. They glowed irregularly. Although we could not examine the effect of the camera flashes on the female receptivity definitely, the female flashes were also recorded. The

Flash parameters	Mean ± SE
Pulse duration	0.10 ± 0.02
Interpulse duration	0.41 ± 0.01
Flash interval	0.51 ± 0.01
Flash rate (per sec)	1.97 ± 0.03

Table 2.
Flash parameters of P. malaccae.

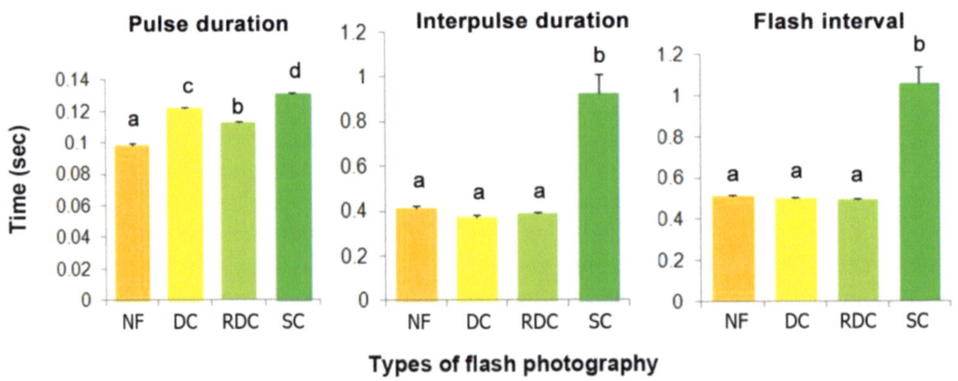

Figure 3.
Comparison of flash parameters (±SE) of P. malaccae after receiving different types of flash photography, no flash or control (NF), digital camera (DC), digital camera with red light autofocusing (RDC), and smartphone camera (SC). Letters indicate significant differences among different types of camera flash treatments.

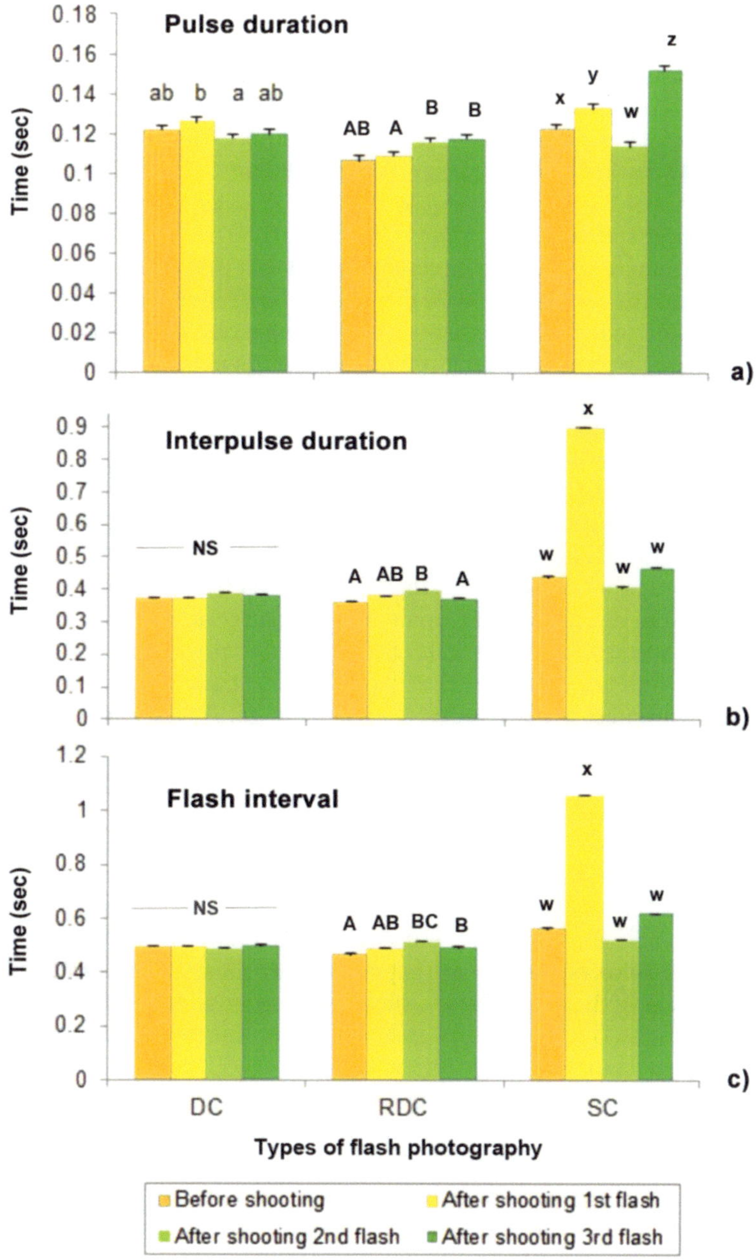

Figure 4.
Comparison of flash parameters (±SE), pulse duration (a), interpulse duration (b) and flash interval (c) of P. malaccae when a set of all camera flash types, digital camera (DC), digital camera with red light autofocusing (RDC), and smartphone camera (SC) is received. Letters indicate significant differences among different sequences of flashing time of each flash photography treatment. NS represents nonsignificant.

illumination from the smartphone caused stopped female flashes with the highest percentage (40%) followed by RDC (20%), while the females in control and DC treatment displayed normal flashing behavior. The observation of oviposition behavior after the shooting flash experiment showed that females from all treatments could lay eggs. Although the females might mate with other males before testing, they can express egg oviposition behavior after receiving flash effect. The fireflies in all flash type experiments can lay eggs for 40–50% with the maximum numbers of eggs, 40–67.

4. Discussion

The experiment of testing camera illumination was carried out using adult *P. malaccae* captured from nature that has no information of age and mating status. Some females probably already mated, and they did not respond to the male flashes and did not show mating posture during the experiment. Although female fireflies could remate after 24 h of first mating examined in *Photinus greeni* [21], there are no reports in *Pteroptyx* species. Thus, it was difficult to interpret the effect of camera illumination on female receptivity. I recommended that preparation of virgin females from culturing for the testing might be the reliable methods to provide the female response signal and mating success data. The reference [22] indicated that the flash patterns of the *Pteroptyx* fireflies in flight and on perch were different. The synchronous flashing on perch acts as attraction signals to make the aggregation groups of the species and might be important as courtship signals. Thus, this study analyzed mainly the signals of synchronous flashes. The adding of females in our experiment helped to induce males to produce the courtship flashes. Most tested males displayed courtship flashes normally. From the results, the flashes likely affect the flash synchronization that is a specific behavior of *Pteroptyx* species. There are many hypotheses of displaying the synchronous behaviors that are (1) attractive communication in dense vegetation environment [22] and (2) male competition to attract a female [23]. It would affect the species communication ability and destroy aggregation ability.

These results indicated that the firefly could adapt their flashing behavior when any strong interferences are encountered. The behavioral plasticity in flashing communication of fireflies was discussed for a long time. Although the flash codes are species specificity, fireflies can change their flash codes depending on the situation and environmental conditions [12, 24]. The plastic behavior might benefit males by increasing their advertisement to females over artificial lighting interference. *Aquatica ficta* male could change their flash signals under different ambient light intensities and different wavelengths of the light [12]. The short wavelength light induced *A. ficta* male flash brighter and slower. Additionally, the firefly could adapt their flashing behavior when they are exposed to the gradient of light (from dim to bright) by increasing their light intensities, while they displayed no flashes when they confronted with the bright light and slightly inhibit the flashing behavior which even appeared in dim environment later.

In the experiment, the tested females did not respond specific flash signals to the males that might cause from mated status of the females or captive condition. Reference [2] reported that the captive *Photuris* female did not respond to any males even the artificial males. The captive status might affect female response in this study.

The camera illuminations produced white light that are a broad spectrum composed of both short and long wavelengths of light. In this study, the response of male fireflies was quite similar to the effect of short wavelength light studied by the researchers [12]. The main biological effect of the light might come from the short wavelength light. All experimental light types were mainly LED flash technology varied with light intensities and function system. The detail of light spectrum was not described. The smartphone illumination was the brightest type that showed the strongest effect when compared with DC and RDC. The fireflies received SC flashes and showed twice the time duration of flash interval than control.

Currently, smartphone photography has become commonly used. The smartphone flashes are considerably developed to improve night shot photographs such as dual flash and Xenon flash that are stronger than LED and might cause adverse impact on the firefly behavior. The peak wavelength of LED and xenon flashes is in the blue region of visible light (400–480 nm) that could adverse biological effects including human eyes and skin when receiving long exposure [25]. The toxic effects of short wavelength light on many insects are known, i.e., mortality in immature stages of *D. melanogaster* [26] and strawberry leaf beetle [27]. In case of fireflies, the short wavelength of light caused flash signal alteration mentioned by the researchers [12]. However, the light from high-pressure sodium street lamps (with peak at low short wavelength) also likely deterred males to locate females; the males preferred to focus on females in darker areas [10]. Similarly, *Pteroptyx* fireflies showed adaptive behavior to avoid receiving the light by staying above the lamp level or staying in darker areas of the trees, even though it might have long-term effects or bring populations down and finally disappear from the habitats. From the field observation of local people in Chao Phraya riverbank area, the *Pteroptyx* fireflies stopped flashing after receiving camera flashes from tourists, moved away from the light side and could not be observed the firefly colonies back for long period up to 3 months. People wondered that the camera illumination might result in firefly mortality. As in the results of the current study, the tested fireflies in the experiment could prove that the temporary receiving of camera illumination did not cause firefly deaths. The life span of the adults is not different from control group. Additionally, the females could display egg oviposition behavior normally.

Timing of receiving flashes is probably an important factor. The tested fireflies were exposed to the camera illumination in the beginning of mating sequences. The 80% of experimental treatment could observe paired fireflies after shooting flashes. It is also possible that the copulation could not continually happen if they are interrupted during the sensitive steps of their mating sequences, i.e., dorsal mouthing. The females might remain for a long time in dorsal mouthing posture as in mating under light pollution of *Sclerotia aquatilis* [11]. Adversely, the females will not allow the male to mate with those that are often found naturally (**Figure 5**).

To conclude, any types of the camera illumination influenced *Pteroptyx* firefly behavior, i.e., mating behavior interruption, courtship signal alteration, synchronous flashing that probably influences female reception. The impact levels depend on illumination types related to light intensity, wavelength, etc. However, the fireflies could have adaptive behavior to avoid the effect after receiving the first trial; however, the repeated exposures are possibly resulting in decreasing numbers and becoming extinct in the natural habitats. Although firefly photography is not good

Figure 5.
The courtship behavior of P. malaccae in nature. The male displayed dorsal mounting on female back and flashed to her eyes. The female might reject the male during the process. Photo: Mr. Banthoon Phankaew.

when using flash illumination, many tourists preferred to use it anyway. Our finding suggests that the impact of understanding would be a useful knowledge for raising public awareness in firefly tourism activity. Besides the effect from camera illumination, the other sources of light should be concerned for firefly conservation.

5. Conclusions

Studies on the effect of artificial night lighting on fireflies have been demonstrated for concerning the effects on firefly populations and their conservation for a decade. However, they have still little understood in particular species and some different types of lights.

The book chapter has presented the impact of artificial light focusing on flash photography from both cameras and smartphones that probably have an impact on *P. malaccae*; the tourism highlighted species in Thailand and also other countries. The results indicated that the flash types could affect their courtship behavior, especially the flash from smartphones or from the brighter sources. The flash photography might bring vulnerability to wild *Pteroptyx* populations especially to the females. The consideration on firefly tourism rules about using camera flashes are needed to protect *P. malaccae* populations in tourism habitats.

Acknowledgements

We thank Sudarat Meephop for assistance in data collection and Dr. Parichart Laksanawimol who helped with the experimental design. This study was supported by Kasetsart University and Bangchak Corporation Public Company Limited.

Conflict of interest

The author declares no conflict of interest.

Author details

Anchana Thancharoen[1*] and Sirima Masoh[2]

1 Department of Entomology, Faculty of Agriculture, Kasetsart University, Bangkok, Thailand

2 Faculty of Science, Prince of Songkla University, Songkhla, Hat-Yai, Thailand

*Address all correspondence to: koybio@gmail.com

References

[1] Mizon B. Light Pollution: Responses and Remedies. London: Springer-Verlag; 2001. p. 219. DOI: 10.1007/978-1-4471-0669-2

[2] Lloyd JE. Stray light, fireflies, and fireflyers. In: Ecological Consequences of Artificial Night Lighting. Washington, DC: Island Press; 2006. pp. 345-364

[3] Luarte T, Bonta CC, Silva-Rodriguez EA, Quijón PA, Miranda C, Farias AA, et al. Light pollution reduces activity, food consumption and growth rates in a sandy beach invertebrate. Environmental Pollution. 2016;**218**:1147-1153. DOI: 10.1016/j.envpol.2016.08.068

[4] Firebaugh A, Haynes KJ. Light pollution may create demographic traps for nocturnal insects. Basic and Applied Ecology. 2019;**34**:118-125. DOI: 10.1016/j.baae.2018.07.005

[5] Hu Z, Hu H, Huang Y. Association between nighttime artificial light pollution and sea turtle nest density along Florida coast: A geospatial study using VIIRS remote sensing data. Environmental Pollution. 2018;**239**: 30-42. DOI: 10.1016/j.envpol.2018.04.021

[6] Manríquez PH, Jara ME, Diaz MI, Quijón PA, Widdicombe S, Pulgar J, et al. Artificial light pollution influences behavioral and physiological traits in a keystone predator species, *Concholepas concholepas*. The Science of the Total Environment. 2019;**661**:543-552. DOI: 10.1016/j.scitotenv.2019.01.157

[7] van Langevelde F, Ettema JA, Donners M, WallisDeVries MF, Groenendijk D. Effect of spectral composition of artificial light on the attraction of moths. Biological Conservation. 2011;**144**(9):2274-2281. DOI: 10.1016/j.biocon.2011.06.004

[8] Eisenbeis G, Hänel A, McDonnell M, Hahs A, Breuste J. Light pollution and the impact of artificial night lighting on insects. In: Ecology of Cities and Towns: A Comparative Approach. New York, USA: Cambridge University Press; 2009. pp. 243-263

[9] Hagen O, Santos RM, Schlindwein MN, Viviani VR. Artificial night lighting reduces firefly (Coleoptera: Lampyridae) occurrence in Sorocaba, Brazil. Advances in Entomology. 2015;**3**(01):24-32

[10] Ineichen S, Rüttimann B. Impact of artificial light on the distribution of the common European glow-worm, *Lampyris noctiluca* (Coleoptera: Lampyridae). The Lamp. 2012;**2**:31-36

[11] Thancharoen A. The biology and mating behavior of an aquatic firefly species, Luciola aquatilis sp. nov. Thancharoen (Coleoptera: Lampyridae). Doctoral dissertation, Ph.D. (thesis), Mahidol University; 2007

[12] Owens ACS, Meyer-Rochow VB, Yang EC. Short- and mid-wavelength artificial light influences the flash signals of *Aquatica ficta* fireflies (Coleoptera: Lampyridae). PLoS One. 2018;**13**(2):e0191576. DOI: 10.1371/journal.pone.0191576

[13] Harasti D, Gladstone W. Does underwater flash photography affect the behaviour, movement and site persistence of seahorses? Journal of Fish Biology. 2013;**83**(5):1344-1353. DOI: 10.1111/jfb.12237

[14] Thancharoen A. Well managed firefly tourism: A good tool for firefly conservation in Thailand. The Lamp. 2012;**2**:130-136

[15] Ballantyne LA. Further revisional studies on the firefly genus *Pteroptyx* Olivier (Coleoptera: Lampyridae:

Luciolinae: Luciolini). Transactions of the American Entomological Society. 1987;**1**:117-170

[16] Ballantyne LA. McLean MR. Revisional studies on the firefly genus *Pteroptyx* Olivier (Coleoptera: Lampyridae: Luciolinae: Luciolini). Transactions of the American Entomological Society (1890-). 1970;**96**(2):223-305

[17] Jusoh W, Ballantyne L, Lambkin CL, Hashim NR, Wahlberg N. The firefly genus *Pteroptyx* Olivier revisited (Coleoptera: Lampyridae: Luciolinae). Zootaxa 2018;**4456**(1):1-71. DIO: 10.11646/zootaxa.4456.1.1

[18] Prasertkul T. Characteristics of *Pteroptyx* firefly congregations in a human dominated habitat. Journal of Insect Behavior. 2018;**31**(4):436-457. DOI: 10.1007/s10905-018-9687-8

[19] Konno J, Hatta-Ohashi Y, Akiyoshi R, Thancharoen A, Silalom S, Sakchoowong W, et al. T i LIA: A software package for image analysis of firefly flash patterns. Ecology and Evolution. 2016;**6**(9):3026-3031. DOI: 10.1002/ece3.2078

[20] Lewis SM, Cratsley CK, Demary K. Mate recognition and choice in *Photinus* fireflies. In: Annales Zoologici Fennici. Finnish Zoological and Botanical Publishing Board; 2004. pp. 809-821. http://www.jstor.org/stable/23736144

[21] South A, Lewis SM. Determinants of reproductive success across sequential episodes of sexual selection in a firefly. Proceedings of the Royal Society of London B: Biological Sciences. 2012;**279**(1741):3201-3208. DOI: 10.1098/rspb.2012.0370

[22] Buck J, Buck E. Biology of synchronous flashing of fireflies. Nature. 1966;**211**:562-564. DOI: 10.1038/211562a0

[23] Buck J, Buck E. Synchronous fireflies. Scientific American. 1976;**234**(5):74-85

[24] Carlson AD, Copeland J. Behavioral plasticity in the flash communication systems of fireflies: Although insect behavior has generally been considered stereotyped, recent research indicates that fireflies can alter their flash patterns according to the behavioral context. American Scientist. 1978;**66**(3):340-346

[25] Arjmandi N, Mortazavi G, Zarei S, Faraz M, Mortazavi SA. Can light emitted from smartphone screens and taking selfies cause premature aging and wrinkles? Journal of Biomedical Physics and Engineering. 2018;**8**(4):447-452. DOI: 10.31661/jbpe.v0i0.599

[26] Shibuya K, Onodera S, Hori M. Toxic wavelength of blue light changes as insects grow. PLoS One. 2018;**13**(6):e0199266. DOI: 10.1371/journal.pone.0199266

[27] Hori M, Suzuki A. Lethal effect of blue light on strawberry leaf beetle, *Galerucella grisescens* (Coleoptera: Chrysomelidae). Scientific Reports. 2017;**7**(1):2694

Chapter 6

Biofluorescence in Terrestrial Animals, with Emphasis on Fireflies: A Review and Field Observation

Ming-Luen Jeng

Abstract

The mysterious world of biofluorescence in terrestrial ecosystems is mesmerizing. Though not as ubiquitous as in the ocean, it is not a rare phenomenon on land. Fluorescence occurs in all major phyla of terrestrial animals (Platyhelminthes, Mollusca, Annelida, Nematoda, Onychophora, Arthropoda, and Chordata) and their subgroups, with diverse fluorophores and performance. In this chapter, we make a general review on the fluorescence in terrestrial animals first, including their systematic distribution, research history, fluorophores, and proposed functions for each group among several other aspects. A systematic observation on the fluorescence of fireflies is reported for the first time. The co-occurrence of biofluorescence and bioluminescence in luminescent land snails, earthworms, potworms, millipedes, and fireflies is a fascinating issue. Though the biochemical mechanism of photogenesis is not fully understood in many terrestrial animals except fireflies, it appears that biofluorescence and bioluminescence do not have clear interaction during the light production process. However, fluorophores and luminophores are usually biochemically related and are different from the photogenic mechanism of jellyfish and several marine creatures whose ultimate light emission is made through energy transfer from bioluminescence to biofluorescence by green fluorescent protein (GFP) or its variants. The role of fluorescence is disputative. In general, nocturnal animals or animals having cryptic living styles, e.g., in earth or under shelters like tree bark or rocks, tend to exhibit UV fluorescence more frequently than animals that are diurnal or inhabit open environments. This pattern is evident in fireflies wherein only nocturnal and luminescent species exhibit noticeable UV fluorescence (likely from luciferin), which is dim or absent in diurnal or crepuscular fireflies. It is unlikely that occasionally induced UV fluorescence in natural environments can play a significant role in intra- or interspecific communication in fireflies or other nocturnal animals.

Keywords: biofluorescence, terrestrial animals, review, fireflies, photogenic organ, nocturnal, luciferin, adaptationism

1. Introduction

Fluorescence is a form of photoluminescence that occurs when a substance emits light caused by absorbing light or electromagnetic radiation [1]. Some amount of the absorbed energy is dissipated in the process, and the rest undergoes an internal energy transition to re-emit light (longer wavelength) almost immediately after the absorption. The substance ceases to fluoresce virtually simultaneously when the stimulating source is removed. A similar but slightly different photoluminescent phenomenon is phosphorescence. It is characterized by slow energy transition, hence a longer emission of light after original excitement. Early scientists tended to use the term phosphorescence to describe all kinds of light emission that did not produce heat (cold light), including bioluminescence (e.g., [2, 3]). The latter is a form of chemiluminescence and should not be mistaken for one another in modern science [4].

European observers reported biological fluorescence (aka biofluorescence) from the sixteenth century [5], but this received significantly less attention when compared to bioluminescence. Although green fluorescent protein (GFP) was discovered and purified in the 1960s [6], it was not until the late 1990s that GFP and its variants suddenly vaulted from obscurity to the limelight, serving as one of the most widely exploited tools in biochemistry and molecular cell biology [7]. Since then, biofluorescence has been cumulatively reported in diverse organisms and microorganisms. It has been found to occur in marine, freshwater, as well as terrestrial ecosystems [5, 8–12]. Interestingly, it appears that biofluorescence is much more common in marine lives and plants than in terrestrial animals [5, 13]. Marine animals fluoresce by absorbing ambient blue or ultraviolet (UV) light in the sea or from their own bioluminescence and glow in cyan, green, orange yellow, or red in human vision. In contrast, terrestrial biofluorescent animals mostly glow under UV illumination, but different light excitations have also been reported [5, 14].

In a broad sense, animals are more or less fluorescent because their cells and tissues contain various endogenous fluorescent substances, such as flavins, reduced NADH and NADPH, lipofuscins, reticulin fibers, collagen, elastin, and chitin, among others [15]. Such autofluorescence in vitro is beyond the scope of the current report; however, it is difficult to separate autofluorescence from externally visible biofluorescence in many circumstances.

In this chapter, we will first review literatures on biofluorescence in terrestrial animals, including systematic distribution, fluorophores, proposed functions, connection with bioluminescence in luminous groups, and applications especially in taxonomic use. We will then move on to the fluorescence in fireflies, giving a historical review of related fluorescence in vitro and introduce a macroscopic observation on externally visible fluorescence in living fireflies.

2. Biofluorescence in terrestrial animals, a review

All major phyla of terrestrial animals, including platyhelminthes, mollusks, annelids, nematodes, onychophorans, arthropods, and chordates, have been shown to have fluorescent species [5, 13], with some taxa better studied than others. The current review covers only truly land-living animals (aka terrestrial), and freshwater and intertidal dwellers have been excluded.

2.1 Platyhelminthes

The epithelium and mucus trail of some land planarians (Tricladida: Terricola) exhibit fluorescence under UV torch or fluorescence microscopy. Yellow and pale brown pigments in integument, likely pheomelanins, give off a dim yellowish fluorescence. It has been suggested that secreted compounds within mucus fluorescence may have repugnatorial or toxic functions [16]. UV-induced crimson fluorescence from rhabdoids was recorded in deparaffinized sections of *Platydemus manokwari* which is indicative of the existence of uroporphyrins. Living specimens, however, did not fluoresce in their rhabdoids [17].

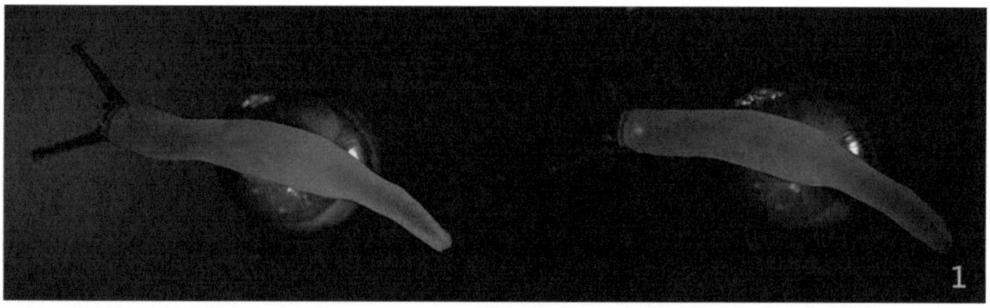

Figure 1.
The glowing land snail Quantula striata *(from Singapore) from ventral aspect under white light (left) and 365 nm UVA torch (right), notice the fluorescent spot of the photogenic organ on suprapedal gland. By Tsan-Rong Chen (CTR).*

Parasitic flatworms have also been reported to be fluorescent in vitro. For instance, *Schistosoma japonicum* (Trematoda: Strigeidida), a blood fluke causing schistosomiasis in humans, can emit a broad spectrum of fluorescence between 500 and 600 nm under different excitation light sources in confocal microscopy. Green (514 nm) and blue (488 nm) lights yielded strong fluorescence of yellowish green (550–580 nm) and are best for microscopic observation [18].

2.2 Mollusca

The land snail *Quantula striata* has been well-known for its bioluminescence [19]. Its eggs and newly hatched snails glow, while immature individuals give off rhythmic flashes in green from a photogenic organ located in the anterior part of the supra pedal gland. A fluorescent substance has been extracted and partially purified from its photogenic organs. The compound exuded green fluorescence (λmax 515 nm) under UV light (365 nm), similar to the snail's bioluminescent spectrum. Dim fluorescence of the photogenic organ in vivo can be excited by UVA light and is visible externally to the naked eye (**Figure 1**). The compound is not water-soluble and shows similar properties to flavins [20, 21].

Fluorescence in land snails may occur either on the shells or the soft body or both. Different fluorescent color patterns have been observed under near UV and near infrared light across different snail families. The color patterns under different illuminations are thought to be environment adaptive [22]. Fluorescent substances from the snails' soft body have been evaluated for systematic and taxonomic application. Using paper chromatography, Kirk et al. compared the characteristic fluorescence and absorption patterns of the extracted substances from seven species of European land snails. The results indicated it is species-specific but were ineffective to differentiate age and geographic or dietary variations within the species [23]. Fluorescent pigments of a sibling species pair of *Bradybaena* snails from Japan were confirmed to be useful in species diagnosis [24].

2.3 Annelida

More than 30 species from five families of earthworms and potworms (subphylum Clitellata: Oligochaeta) are known to be luminescent. Some of them, such as the cosmopolitan *Microscolex phosphoreus*, North American *Diplocardia longa* (both in Haplotaxida and Acanthodrilidae), and several European *Eisenia* species (Lumbricidae) have been extensively studied for their bioluminescence [25–31]. Two types of bioluminescence with different chemical, physical, and biological features were detected in earthworms [30]. Most of the luminescent species have flavin-derivative bioluminescent substances stored in granule-filled coelomic mucocytes. The light production occurs when coelomic fluid (mucus) is discharged. However, the body cavity of earthworms only becomes luminous while dying because their coelomocytes break up inside the body. Oppositely, Siberian *Fridericia heliota* and an unidentified *Henlea* species (both in Enchytraeidae) can glow from body walls, whereas the former does not even have glowing mucus. The luminophore in the luciferin of *F. heliota* has been determined to be a tyrosine derivative. All luminescent substances known so far are also fluorescent under UV light and have similar emission spectrums with those of bioluminescence [30].

Several flavin derivatives have been isolated from both luminescent and non-luminescent earthworms, wherein riboflavin (vitamin B2) in the unbounded state was found in their coelomocytes [32]. The coelomic fluid of luminescent *Eisenia lucens* fluoresces yellow-green light initially and turns into blue when bioluminescence ceases. In contrast, fluorescent color changing did not occur in nonluminescent *E. fetida.* It suggests that some product of the luminescence reaction changes the color of fluorescence. It is thus postulated that the incapability of bioluminescence in nonluminescent earthworms is due to their lack of a certain component to convert riboflavin into lumiflavin in the oxidative system [29].

Riboflavin exists in the cytoplasm of coelomocytes with different contents among earthworm species of various genera [33]. The secreted mucus from intersegmental pores and mouths also produces fluorescence [29, 34, 35]. Since riboflavin is essential to the regeneration, the stem cells responsible for regeneration accumulate more riboflavin than other cells thus inducing stronger autofluorescence [36].

The coelomic fluid also yields species-specific intensity of fluorescence. For example, *Eisenia andrei* and *E. fetida*, which are hard to differentiate morphologically, each carries a specific fluorescence fingerprint in the cell-free coelomic fluid [37]. A broader taxon examination of the coelomocyte-derived fluorescence proved its value in supravital species identification of morphologically resembling earthworms (**Figure 2**) [38].

2.4 Nematoda

Living nematodes from 15 selected genera have been found exhibiting pale yellow to green fluorescence from their intestines, spicules, and lips in microscopic observation under both blue (450–490 nm) and UV (365 nm) epi-illumination [14]. Aging nematodes tend to have stronger fluorescence than juvenile individuals. The intensity of fluorescence could be an indicator of age and viability of nematodes [14]. When the nematodes were dying or killed by physical or chemical treatment, blue fluorescence burst from the intestine into the entire body in a wave. This dramatic phenomenon is known as "death fluorescence" [39]. The blue fluorescent substance was inferred to be lipofuscin, which is stored in intestinal lysosome-related

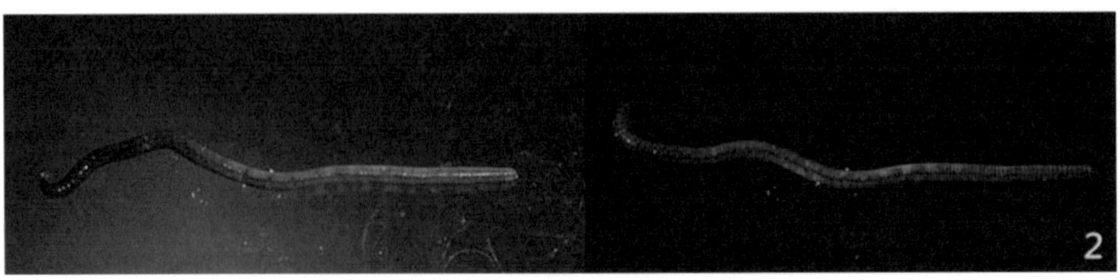

Figure 2.
A nonluminescent earthworm (from Taiwan) emitting yellow-green fluorescence (right) under 365 nm UVA torch. By CTR.

organelles and is cumulative with age [14]. Later research identified anthranilic acid glucosyl esters (derived from tryptophan, not lipofuscin) to be the source of death fluorescence [39]. This revealed that organismal death of nematodes has a similar mechanism to necrotic propagation in mammals [39].

2.5 Onychophora

Research on the fluorescence of velvet worms is limited. Weak cyan fluorescence in vitro was seen on the outer zone of the cuticle specimens embedded in frozen sections of South African *Peripatopsis moseleyi* (Euonychophora: Peripatopsidae), but not on claws or jaws [40]. In a previous work on the same species, however, neither its cuticles nor fresh slime expelled fluorescing [41].

2.6 Arthropoda

Land-dwelling arthropods are found in all four subphyla: Chelicerata, Myriapoda, Hexapoda, and Crustacea (paraphyletic). A wide array of terrestrial arthropods has been reported fluorescing upon UV excitation [41–49]. No fluorescent land crustacean has yet been reported, though crab eye lenses are known to be fluorescent.

2.6.1 Chelicerata

Chelicerates are featured by the absence of antennae and jaws and the presence of chelicerae as their feeding appendages. Terrestrial groups are exclusively in class Arachnida and consist of mites (order Acariformes), ticks (Parasitiformes), harvestmen (Opiliones), camel spiders (Solifugae), hooded tickspiders (Ricinulei), pseudoscorpions (Pseudoscorpiones), scorpions (Scorpiones), whip scorpions (Uropygi), tailless whip scorpions (Amblypygi), and spiders (Araneae). Fluorescent species have been recorded in at least eight out of the 10 orders (**Figure 3**), and more are awaiting future research [41–45].

Two fluorescent sources are known in chelicerates: excitation of cuticular fluorophores like beta-carboline and coumarin in hyaline layer, and integumentary/hemolymph fluorophores like tyrosol [5, 43, 50]. The former occurs only in extant scorpions and horseshoe crabs and extinct sea scorpions (Eurypterida), whereas the latter is prevalent in the other chelicerates which lack the hyaline layer.

Figure 3.
Biofluorescence in chelicerates. (a) bark scorpion Lychas scutilus *(Scorpiones: Buthidae); (b) harvestmen (Opiliones); (c) trapdoor spider* Liphistius malayanus *(Megalomorphae: Liphistiidae); (d)* Gnathopalystes *huntsman spider (Araneomorphae: Sparassidae); (e)* Gasteracantha *spiny orb-weaver spider (Araneomorphae: Araneidae). All animals from Singapore, under flash light/365 nm UVA torch. By Nicky Bay.*

The cuticular fluorescence is much stronger than hemolymph one and remains unfaded for years in museum specimens. The presence of the hyaline layer and accompanied UV fluorescence has been suggested to be a plesiomorphic (ancestral) trait in Chelicerata [43]. In this scenario, UV fluorescence either has no adaptive significance for scorpions or just serves as a function common to scorpions, horseshoe crabs, and eurypterids. Therefore, it might not be an adaptation of scorpions for land living [43, 50]. Alternatively, UV fluorescence in scorpions has evolved a new adaptation deviated from its initial role in their marine ancestors and relatives. For example, body fluorescence serving as a shelter-finding indicator has been proposed [51]. Recently several species of *Chaerilus* scorpions (Chaerilidae) were discovered to be non UV fluorescent and independent of their various living environments (in soil, epigean, or cave-dwelling) [52]. This exception provides an insight into the long-debated role of fluorescence in scorpions. The loss of fluorescence is surely an evolution novelty in scorpions, but its irrelevance to habitat types rendered the adaptationist explanation questionable (see [53]).

UV fluorescence is widespread across families among spiders [41, 45, 54]. There are two suborders of spiders: Mygalomorphae (e.g., tarantulas) and Araneomorphae (modern spiders). Mygalomorphae resembles the ancient form in morphology: covered with tergites on the dorsum and fluoresces only from intersegmental membranes and appendage tips [54] (**Figure 3**). On the other hand, fluorescence occurs in many araneomorph families, especially in the highly derived Entelegynae [45, 54] (**Figure 3**) family. Most araneomorph spiders have lost their tergites, permitting UV light to penetrate the integument, triggering hemolymph fluorescence [51]. As far as is known, only a few species fluoresce through the entire body. Among body parts, fluorescence occurs most commonly in eyes and joints, moderately in abdominal hemolymph, and least in the cephalothorax. It appears that all araneomorph spiders fluoresce to various degrees and each species has its species-specific emission of fluorescence [45]. Spider eggs of both suborders examined thus far are all fluorescent. Egg sacs also fluoresce, though common but not universally [54]. Web silk emits weak yellow-green fluorescence under UV light [55].

UV fluorescence has evolved repeatedly in spiders' evolution and is liable to keep changing among species. Fluorescent patterns and fluorophores also differ among families [45]. Its function in spiders, however, is unclear. The evolution of fluorescence has been postulated to be driven by pray-predator interaction, sexual selection, or photoprotection [56–60]. Most of the spiders have poor vision and largely rely on mechanical and chemical cues for predation and mating. Fluorescence is unlikely to function as intraspecific signals. Instead, fluorescence may help spiders blend into ambient environment and thus reduce perception by their prey and predators [45, 56]. Jumping spiders (family Salticidae) have excellent vision. Many species show sexual dichromatism, and these males have elaborate mating dances. Some marking patches in males carry dancing-relevant fluorescent signals. Experiments provided convincing evidence that the dynamic fluorescent signals play a vital role for jumping spiders' mate choice [58, 59]. Another postulated function for UV fluorescence in spiders is for protection from UV radiation [60]. A similar hypothesis had been proposed earlier for the evolutionary origin of fluorescence proteins in corals [61]. A transcriptome analysis of spiders has suggested that they are unable to synthesize melanin, a common light-absorption pigment almost universally present in all organisms [62]. Melanin in insects and crustaceans operates additionally as an innate immune system [63]. It appears compelling that UV fluorescence is a photoprotective mechanism for araneomorph spiders which do not have melanin and tergite. Recently, however, melanin has been confirmed to be present in spiders, thus diminishing the validity of this hypothesis [64].

2.6.2 Myriapoda

Myriapoda comprises four orders: Chilopoda (centipedes), Diplopoda (millipedes), Symphyla, and Pauropoda. UV fluorescence is currently known only in the first two taxa [5, 41, 46, 47, 65]. Several centipedes and millipedes are luminescent [26, 46, 47, 66–68]. This raises the question of whether fluorescence and luminescence in myriapods are a mechanistic link for the production of light, whereby the fluorescent substance is the ultimate light emitter through energy transfer as in GFP [7, 46].

The sublittoral centipede *Orphaneous brevilabiatus* (Geophilomorpha: Oryidae) can discharge yellowish bioluminescent slime while walking, leaving a shiny trace with a fruit odor lasting from a few seconds to about 2 minutes [66, 67]. The glowing slime is secreted from coxal glands, and the light results from a luciferin-luciferase interaction. The slime does not show fluorescence [66]. The other glowing centipedes are found in five families, mostly in order Geophilomorpha, and share more or less similar bioluminescent behaviors, with variations in slime colors or luminescent intensity [67]. The slime is apparently not for intraspecific visual signaling since geophilomorph centipedes are eyeless. Rather it is a defensive mechanism against predators [67]. Weak UV fluorescence has been reported to occur in some nonluminescent *Cormocephalus* species (Scolopendromorpha: Scolopendridae) [41].

In millipedes, only some 10 out of the 12,000 described species are luminescent [47, 67]. Co-occurrence of photoluminescence and chemiluminescence has been documented in North American *Motyxia* species (also known as Sierra luminous millipedes, in Xystodesmidae) but is unclear for the other glowing millipedes (distribution in Asia and Pacifics, e.g., *Spirobolellus* in Spirobolidae, and *Salpidobolus* (*Dinematocricus* in most references) in Rhinocricidae). *Motyxia* millipedes glow spontaneously or by physical stimulation, with a bright greenish-white hue throughout the entire body [67]. A bioluminescent substance was isolated from *M. sequpiae* and identified to be 7,8-dihydropterin-6-carboxylic acid [69]. This compound is unstable outside of the cuticle, leading to pterin-6-carboxylic acid which is also found in the cuticle of *M. sequoiae*. Both compounds are UV fluorescent, showing emission peak at 505 and 450 nm, respectively. The emission spectrum of 7,8-dihydropterin-6-carboxylic acid in vivo and in vitro is very close to the bioluminescence of *M. sequoiae* (peaked at 495 nm), and Kuse et al. suggested it as the light emitter [65]. Pterin-6-carboxylic acid was later found in a non-bioluminescent xystodesmid, the Japanese train millipede *Parafontaria laminata armigera*. It fluoresces upon direct UV excitation and gives off a blue emission [65, 68]. Autofluorescence is widespread in diplopod orders like Spirobolida, Siphonophorida, and Polydesmida (**Figure 4**) [46].

Based on the evidence so far, the mechanisms of light production in glowing jellyfish and *Motyxia* millipedes seem different. The chemiluminescence in the latter produces a longer wavelength of light than that of fluorescence but reversely so in the former. Though the luminophore and fluorophore in *Motyxia* are biochemically related, their mechanistic interaction in photogenesis is unclear (see Section 3.1).

Bioluminescence in millipedes has been demonstrated to be an aposematic signal. Glowing fake millipedes were found to have a much lower predation rate than non-glowing ones in a field experiment [70]. It was postulated that the evolution of luminescence in *Motyxia* may have initially been triggered by a harsh environment to deal with metabolic stress and then was later repurposed for aposematism [47]. Biofluorescence, on the other hand, appears to play an insignificant role ecologically or ethologically, if any at all [5].

Figure 4.
A non-luminescent millipede (from Singapore) fluorescing in blue hue under flash light/365 nm UVA torch. By Nicky Bay.

2.6.3 Insecta

Insects are the most diverse animals, not only in terms species richness but also in morphology, physiology, behaviors, and ecological niches. Bioluminescence has evolved in some beetle families (Elateridae, Lampyridae, and Rhagophthalmidae), flies (in Keroplatidae or Mycetophilidae s. lat.), and springtails (order Collembola, six-legged arthropods allied to insects) [26, 71]. South American roaches *Lucihormetica* (Dictyoptera: Blaberidae) were once reported to be luminescent [72] but are actually fluorescent [73, 74].

Fluorescence seems ubiquitous in most, if not all, insect orders and occurs not only in the insect body but also in their eggs, egg cases, silks, exuviae and other products (**Figure 5**) [41, 48, 49, 73, 75]. Fluorescent materials include pterin, flavin, and kynurenine derivatives, chitin, resilin, and luciferin (thiazole derivative), among many others [5, 48, 76–80]. Since chitin is autofluorescent, insects with weakly sclerotized cuticles tend to show stronger fluorescence than those that are heavily armored-like beetles [49]. Eyes, markings on body or wings (usually white, yellow, or cyan in color), and joints are the most frequent fluorescent parts [41, 48, 81–84]. Fluorescence in fireflies will be addressed in the next section.

Fluorescence may play roles in intra- and interspecific communication as suggested in butterflies, moths, damselflies, and dragonflies [48, 82, 83, 85], but it is premature to make a conclusion about their function at present [85]. Autofluorescence of resilin, chitin, and some other substances could just be an epiphenomenon and have no adaptation value.

2.7 Chordata

Traditionally, chordates constitute three subordinate groups: lancelets (Cephalochordata), tunicates (Tunicata), and vertebrates (Vertebrata). Terrestrial forms are exclusively vertebrates. UV fluorescence in land vertebrates was first noted in some plumage areas of Australian parrots and later found prevalent in most parrot species worldwide as well as in some other bird species [5, 85–88]. A few fluorescent amphibians (South American tree frog *Hypsiboas punctatus* [Anura: Hylidae]), reptiles (African and Madagascan chameleons [Squamata: Chamaeleonidae]), and mammals (North America flying squirrels *Glaucomys* species [Rodentia: Sciuridae]) were discovered recently [13, 89, 90]. Mouse skin emits red fluorescence that peaks at 674 nm, and the fluorophore is found to be derived from food [91]. In addition, autofluorescent substances like keratin, collagen, and enamel, among others [15],

Figure 5.
Biofluorescence in various insects or their products. (a) stick insect Necroscia punctata *(Phasmatodea: Heteronemiidae); (b) long-horned grasshopper (Orthoptera: Tettigoniidae); (c) true bug's nymphs and eggshells (Hemiptera: Pentatomidae); (d)* Paralecanium *scale insect (Hemiptera: Coccidae); (e)* Aspidomorpha *tortoise beetle (Coleoptera: Chrysomelidae); (f)* Cerosterna *long-horned beetle (Coleoptera: Cerambycidae); (g) larvae of* Macrolycus *netwinged beetle (Coleoptera: Lycidae); (h) Red slug caterpillar* Eterusia aedea *(Lepidoptera: Zygaenidae); (i) cocoon of a lichen moth (Lepidoptera: Erebidae). (a-g) from Singapore, under flash light/365 nm UVA torch, by Nicky Bay; (h-i) from Taiwan, under white light/395 nm near-UVA torch, by JML.*

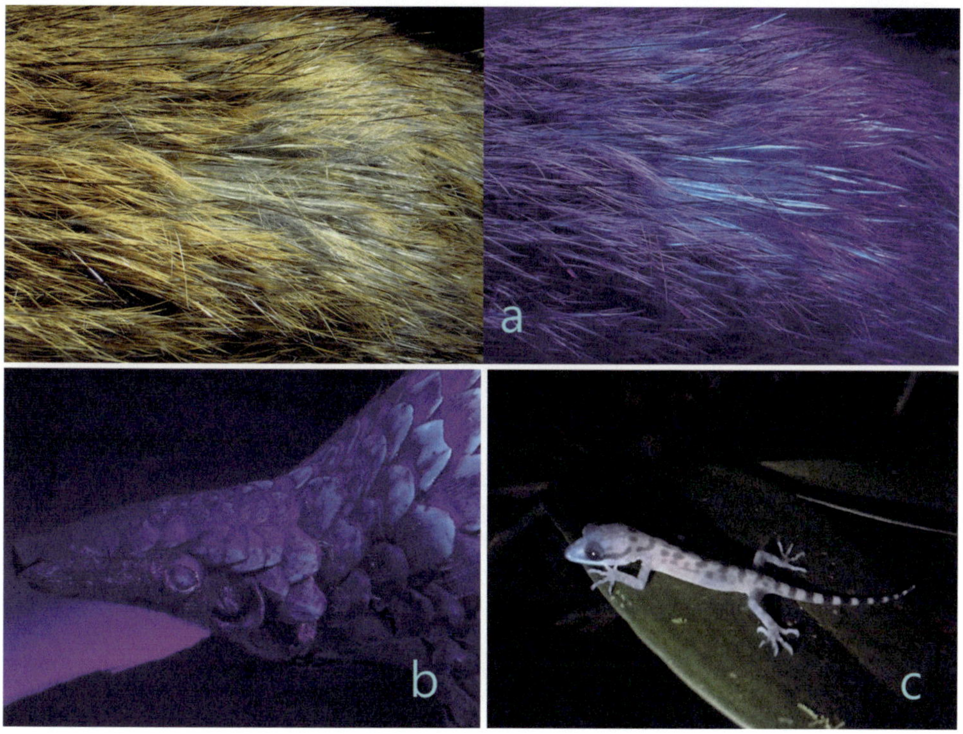

Figure 6.
Biofluorescence in mammals and reptile. (a) Coxing's white bellied rat Niviventer coninga *(Rodentia: Muridae), with fluorescent setae. (b) Chinese pangolin* Manis pentadactyla *(Pholidota: Manidae), with fluorescing keratin scales. (c) Kinabalu bow-fingered gecko* Cyrtodactylus baluensis *(Squamata: Gekkonidae), with bone-based fluorescence. (a-b) Taiwan specimens in NMNS collection, under white light/395 nm near-UVA torch; (c) from Sabah, Malaysia, under 365 nm torch. By JML.*

also lead to UV fluorescence in vertebrate claws, nails, hairs, scales, skin, teeth, and bones to various extents (**Figure 6**). Known vertebrate fluorophores include fulvins, carotenoids, spheniscin, hyloins, pheophorbides, etc. [5, 13, 91].

Intraspecific communication through UV fluorescent signals has been proposed in parrots, frogs, and chameleons [13, 85, 87, 89, 92]. Under natural illumination, fluorescence found on the yellow crown of the budgerigar *Melopsittacus undulatus* (Psittaciformes: Psittaculidae) may enhance visual perception. Both male and female parrots prefer to be associated with fluorescent potential mates over nonfluorescent ones which had been treated with sunblock gel [87]. Pearn et al. [93], however, did not note any fluorescence-linked difference in mate choice. In the frog case, Taboada et al. demonstrated that the emission spectrum of fluorescence from lymph and skins matches the sensitivity of night vision in amphibians and considerably enhances brightness of the individuals under twilight and nocturnal scenarios. This may increase their detectability by conspecifics while remaining cryptic to their predators in a dusk environment [13]. In regard to chameleons, they usually live in more closed habitats like forests which have a higher relative component of ambient UV light [94]. They fluoresce in blue color (around 430 nm) which contrasts well with the green and brown background reflectance of forests. Their fluorescent bony tubercles are sexual dimorphic and may play some roles in sexual selection [89]. Stuart-Fox et al. showed that display colors of chameleons occupying more shaded environments have a relatively higher UV component [95].

3. Biofluorescence of fireflies

Researches of fluorescence are extremely scant in comparison to bioluminescence of fireflies in all scientific fields. Most were infrequent studies done in the twentieth century and made only limited progress. We will make an extensive review of related fluorescent substances first and then report our field observations of externally visible fluorescence in fireflies.

3.1 Biofluorescence in vitro, a historical review

Similar to annelids and millipedes, photoluminescence coexists with chemiluminescence in luminous firefly species wherein their luminophores are simultaneously fluorophores (e.g., oxyluciferin). Other fluorescent substrates, like resilin and chitin, are widespread in Lampyridae and irrelevant to photogenesis.

A firefly-related fluorescent substance was extracted from the thoracic and abdominal photogenic organs of a click beetle species (Elateridae: *Pyrophorus noctilucus*) from West Indies and was named pyrophorine by Dubois in the late nineteenth century [96, 97]. It showed similar chemical and physical properties with those of esculin, a glucose compound, and did not interact with luciferase to produce light. Dubois suggested that pyrophorine may have a resonance property which can transform the absorbed invisible rays into visible light, thus intensifying the animal's luminescence which he called "condensed light" [96]. Similar ideas were welcomed by contemporary scientists, but Dubois disagreed on solar radiation being the energy source of condensed light as was commonly thought [100]. The idea was falsified by Coblentz who demonstrated nonoverlapping spectra of pyrophorine fluorescence and firefly luminescence [97, 98]. Dubois persisted his hypothesis of "condensed light" [99].

A similar substance was later extracted from North American fireflies and was named luciferesceine. McDermott regarded it an incidental material irrelevant to photogenesis in fireflies since the substance was also found in nonluminescent species [100]. Dubois agreed to use luciferesceine consistently rather than pyrophorine [99]. Unfortunately the identity of luciferesceine or pyrophorine remained unknown until now, though a pteridine with ribityl residue was inferred [76, 80]. Another fluorescent substance was extracted and purified from the Japanese firefly *Luciola cruciata* [80]. Its property is markedly different from luciferesceine. It was determined to be 8-methyl-2,4,7(1H,3H,8H)-pteridinetrione, a pteridine named luciopterin [76, 80].

Firefly luciferin, commonly known as D-luciferin, is a thiazole derivative ($C_{11}H_8N_2O_3S_2$) [76, 80, 101, 102]. It is fluorescent as a crystalline form or in aqueous solution and shifted its excitation peak from 327 to 395 nm in strong basic solution without changing emission peak at 530 nm [102]. The ultimate product of photogenesis, oxyluciferin, is the real light emitter in firefly bioluminescence [103–107]. Oxyluciferin is extremely unstable, and scientists have used several analogues to study its biochemical and photochemical properties and photogenic mechanism. Their tautomers give different fluorescent emission of spectrum in different pH environments. Keto-form oxyluciferin fluoresces in red in vitro ($\lambda max \sim 635$ nm) but turns into yellow green ($\lambda max \sim 530$ nm) with a blue shift when binding with luciferase in reaction. The latter is identical to the emission of the bioluminescent spectrum. It is postulated that the intermolecular interactions and polarity affect the color of emission of oxyluciferin [106, 107]. Shimomura indicated that the fluorescence spectrum measured after the light emission, even for only a few milliseconds, cannot be considered the fluorescence spectrum of the light emitter [104].

Smalley et al. noted that there were two fluorescent compounds localized in the firefly photogenic organs, one in the photocyte granules and the other in the dorsal layer of the lanterns [108]. The former fluoresces dimly with a cyan-green emission (λmax between 510 and 540 nm) and glows brightly when in basic solution. The latter emits a cyan fluorescence (λmax between 510 and 520 nm) which disappears quickly when exposed to water. The former is likely luciferin but the latter remains unclear.

A red fluorescent material isolated from *Photinus pyralis* was reported by Metcalf in the earth in the 1940s but not identified [109]. Fluorescent pigments in paper chromatography were popularly applied in invertebrate taxonomy as a diagnostic character in species and genus in the 1950s–1960s [23, 77]. Wilkerson and Lloyd detected 51 fluorescent compounds by paper chromatography from 13 North American firefly species of 5 genera (*Photinus*, *Micronaspis*, *Pyrectomena*, *Pyropyga*, and *Photuris*). Each genus did have its own specific pigment contents. [110].

3.2 Biofluorescence in vivo, a macroscopic observation

Not much externally visible fluorescence in fireflies has been documented. In addition to Smalley and her colleagues' findings [108], we found only two related studies [111, 112].

Blue fluorescence resulting from resilin in lenses of compound eyes of *P. pyralis* was reported in 1970 [111]. Actually this protein is common and widespread in arthropod eye lenses [113]. This matches our observation that compound eyes are among the most frequently fluorescing body parts of fireflies under UV illumination, giving a cyan or blue emission.

Recently Yiu and Jeng reported an interesting case of fluorescence in a paedomorphic (neotenic or "larviform") female *Oculogryphus* firefly in Hong Kong [112]. The female glowed in yellow-green light from a pair of photogenic organs in the abdomen and emitted cyan fluorescence through the whole body under UV illumination. This combination is unique to terrestrial animals, even in luminescent groups. It appears that the two systems are mechanistically independent and share some commonness with millipedes in having a shorter wavelength of light emission in biofluorescence than in bioluminescence.

This case triggered our curiosity to explore the secret world of biofluorescence in fireflies. By using UVA or near-UV torches (λmax 365 or 395 nm) in the field and laboratory, we demonstrated that UV fluorescence is widespread across most of the luminescent firefly groups in East and Southeast Asia (**Figures 7** and **8**). In addition to blue fluorescence from resilin in eye lenses, photogenic organs consistently emit cyan fluorescence to naked eyes, no matter if they are glowing/flashing or not. Fluorescence is brighter when the photogenic organs are glowing (**Figure 7a–j**). The fluorescent substance in photogenic organs is likely luciferin as previously suggested [108]. Alcoholic or dry specimens have much weaker fluorescence in their photogenic organs (e.g., **Figure 7k**). The other body parts frequently fluorescing include head capsule (blue emission probably by resilin as seen in eye lenses), intersegmental membrane, and weakly sclerotized cuticles in both adults and larvae (pale blue emission likely autofluorescence of chitin or cuticular proteins) [49]. In contrast, diurnal genera like some *Pyrocoelia* (part), *Vesta*, *Lucidina*, *Pristolycus*, and *Drilaster* species that carry conspicuous or highly contrasting coloration do not display noticeable UV fluorescence except in vestigial photogenic organs if they have them (**Figure 7l–n**).

Figure 7.
Fireflies under white light/ UVA illumination. (a-b) Abscondita cerata *(T), m &f; (c)* Aquatica leii *(C), m; (d)* Asymmetricata ovalis *(M), m; (e)* Pygoluciola *sp. (C), m; (f)* Pyrophanes *sp. (M), m; (g)* Triangulara frontoflava *(M), m; (h)* Sclerotia substriata *(M), m; (i)* Diaphanes *sp. (M), m; (j)* Pyrocoelia bicolor *(M), m; (k)* Lamprigera tenebrosa *(I), m specimen; (l)* Pyrocoelia *sp. (C), m specimen; (m)* Vesta saturnalis *(C), m; (n)* Pristolycus kanoi *(T), m specimen; C=China, I=India, J=Japan, M=Myanmar, T=Taiwan; m=male, f=female. Taxonomically, (a-h) and (n) of Luciolinae, (i-m) of Lampyrinae. Ecologically, all but (l-n) are nocturnal. Under 365 nm or 395 nm near-UVA torch. By CTR and JML.*

Paedomorphic female fireflies are inferred to produce fluorescence throughout the whole body as in *Oculogryphus* because their soft cuticles constitute rich cuticular proteins which are autofluorescent [49, 108]. This is generally true, but the fluorescence is quite dim and neglectable compared to the fluorescence by photogenic organs. For example, *Diaphanes* and *Pyrocoelia* females which show moderate paedomorphism (**Figure 8a–b**) exhibit weak fluorescence through translucent cuticles. Highly paedomorphic females of *Rhagophthalmus* glowworms (Rhagophthalmidae) have a slightly brighter fluorescence (**Figure 8c**). The strong cyan fluorescence of *Oculogryphus* females appears to be an exception, though the paedomorphic degree of the females is approximate to that of *Lamprigera* and *Rhagophthalmus*. The fluorescence of *Oculogryphus* females likely results from luciferin dispersed in the whole body since they can glow bodywide. Some closely allied genera like *Stenocladius* and *Brachypterodrilus*, among others, which are capable of glowing through the whole body [114, 115], may be able to fluoresce in the same manner.

Interestingly, our team documented fluorescent pygopods in one peculiar *Pygoluciola* larva (**Figure 8d**). Pygopod is a pair of elastic and extractable suckers, serving locomotion and cleaning purposes in firefly larvae [115]. The said *Pygoluciola* larva was found near the water's edge in a mountain creek and fluoresced in a blue light under UV from its pygopods and intersegmental membranes. Surprisingly, fluids secreted by the pygopods were fluorescent as well. We examined several earth-living larvae of *Pyrocoelia* and *Diaphanes* fireflies in nearby environments and did not find a similar phenomenon.

Eggs still within females' body of a *Lamprigera* species in Taiwan exhibit vivid yellow-green fluorescence (**Figure 8e**). Its fluorescent substance is unclear since the light emission is noticeably different to the naked eye from that of luciferin, chitin, resilin, and other proteins.

Figure 8.
Paedomorphic females and larva of fireflies under white light/ UVA illumination. (a) Diaphanes *sp. (Myanmar); (b)* Pyrocoelia atripennis *(Japan); (c)* Rhagophthalmus jenniferae *(Taiwan); (d)* Pygoluciola *sp. (China), larva; (e)* Lamprigera yunnana *(Taiwan). Taxonomically, (a-b) and (e) of Lampyrinae, (c) of Rhagophthalmidae, (d) of Luciolinae. Ecologically, all are nocturnal. Notice the blue fluorescence of head and eyes in (b), and pygapod in (d); yellow green fluorescence of eggs inside female's body in (e). Under 365 nm or 395 nm near-UVA torch. By CTR and JML.*

3.3 Adaptive or not?

Noticeable UV fluorescence only occurs in luminescent fireflies and is very dim or nearly absent in diurnal and crepuscular species. Therefore, we have restricted the following discussion about the function of fluorescence to nocturnal fireflies only.

With few exceptions, luminescent fireflies are active from twilight to deep night [112, 115, 116]. During 1.3–2 hours from sunset to total darkness, the ambient light spectrum changes from long wavelength light (orange-red) dominant at sunset, to a mix of shorter wavelength light from sky (blue) and longer wavelength light from vegetation reflection (green) in twilight, to very dim blue light in early darkness, and ultimately to a dominant long wavelength light (moonless) or neutral (full moon) but low intensity at night [94, 116, 117]. In theory UV fluorescence of luminescent fireflies can occur only in twilight and early night when there is more ambient shorter wavelength light. However, we never see any fluorescing firefly with the naked eye in the field. Naturally induced UV fluorescence may happen very occasionally and is thus hard to be a reliable signal to learn or to perceive. In addition, firefly bioluminescence is much brighter and serves the duel purposes of courtship and aposematic signals. Relatively weak fluorescence is redundant as an extra signal system [112]. Furthermore, fluorescence carries no species-specific variation and thus has lower efficacy as a signal than bioluminescence does. This might also be true for firefly larvae which are night hunters and always display bioluminescence aposematism [118].

Can fireflies detect their own UV fluorescence? Recent studies provided inspiring but inconclusive cues. Most of the beetle families have three-opsin color visual system, corresponding to UV, blue wavelength, and long wavelength light (LW), respectively [119]. Transcriptome and phylogenetic analyses revealed that fireflies may have lost blue-sensitive opsin since divergence from the last common ancestor of the family [119, 120, 121]. Adult fireflies keep two visual sensitivity peaks: yellow-green light by LW opsin (λmax 550–580 nm) and near-UV light by UV opsin (λmax 360–420 nm) [121, 122]. As a result, fireflies would be insensitive to the most common emission spectra of blue and cyan fluorescence. However, several studies demonstrated that fireflies did respond to manipulation of blue light, changing their preference or flashing frequency [123, 124]. More studies are required in order to solve the discordance between genomics and ethology.

4. Conclusion

Biofluorescence occurs in all major land animal phyla and subgroups, with diverse fluorophores and performance. Co-occurrence of bioluminescence and biofluorescence is an interesting phenomenon existing in both marine and land animals and calls for more investigation. The GFP-like mechanism of fluorescence, however, has not been found in terrestrial luminescent animals. The latter emit a shorter or subequal wavelength of biofluorescence than that of bioluminescence like in glowing earthworms, millipedes, and fireflies but in reverse in the former [30, 65, 68, 102]. The light emission of luminescent land animals is the result of luciferase-luciferin interaction. The fluorescent substance may be biochemically related to the luminescent material, but does not contribute to the photogenesis given the evidence thus far [29, 65, 102–104].

The role of biofluorescence is disputed. Although some solid cases of birds and spiders are supported by experiment manipulations, most postulated functions require further investigation. The new findings from nonfluorescent *Chaerilus* scorpions are worth a particular mention [52]. It provides a critical and inspiring example to reexamine the role of scorpions' fluorescence. The study of horseshoe crabs and the long-extinct sea scorpions [43] suggested that fluorescence is most likely an old trait that initially evolved in distant ancestors living in the sea and remained in most of the extant descends and is not necessarily an adaptation for land living scorpions. It is suggested that any assertion of adaptation should be considered under a phylogenetic framework [54]. Exceptional cases in the same or closely related groups may provide a good chance to test the postulated advantage.

Empirically, nocturnal animals or animals having cryptic living styles tend to exhibit UV fluorescence more frequently than animals that are diurnal or live in open environments (see [95]). This trend is clear in fireflies wherein only nocturnal and luminescent species exhibit noticeable UV fluorescence other than autofluorescence by chitin, resilin, cuticular proteins, etc. In contrast, diurnal or crepuscular fireflies, though having greater opportunities to be exposed to UV excitation, do not display fluorescence. It makes little sense that seldom induced UV fluorescence, if any, in the daytime or night, can serve as an efficient or reliable communication means for fireflies, either intra- or interspecifically.

It is presently premature to assert the role of fluorescence for nocturnal animals. We argue functionless should be the null hypothesis as classic scientific approach suggests. Further evidence from visual perception, physiological response, and particularly behavioral assay in both lab and field should be collected to test the hypothesis.

Acknowledgements

We are grateful to our colleagues and friends who helped during field trips to China, Southeast Asia, and Taiwan. They are Chih-Yuan Chang, Tsan-Rong Chen, Hui-Yun Tseng, Jing-Fu Tsai, Yen-Jean Chen, and Hai-Ning Wang, Yu-Tang Wang, Chang-Ching Chen, Hua-Te Fang (Taiwan), Jin-Teng Zhao (China), Hilario Cahilog and Reagan Joseph T Villanueva (Philippines), and Kyaw Swar (Myanmar). Special thanks are due to Ai-Meng Ma (aka Nicky Bay) who provided us with excellent biofluorescence images. Technique and equipment for UV photographing and more related photos can be found in his blog: https://www.nickybay.com/2016/02/alien-transformations-under-ultraviolet.html.

The study was financially supported by the Ministry of Science and Technology, Taiwan, grant no. MOST102-2313-B-178-003-MY3.

Conflict of interest

The authors declare that there is no conflict of interest.

Author details

Ming-Luen Jeng
National Museum of Natural Science, Taichung City, Taiwan

*Address all correspondence to: jengml@nmns.edu.tw

References

[1] Rost FWD. Fluorescence: Physics and chemistry. In: Rost FWD, editor. Fluorescence Microscopy. Vol. 1. Cambridge: Cambridge University Press; 1992. pp. 11-12

[2] Watasé S. On the Physical Basis of Animal Phosphorescence. Boston: Ginn & Company; 1896. 118 p

[3] Nutting CC. The utility of phosphorescence in deep-sea animals. The American Naturalist. 1899;**23**:793-799

[4] Siraj N, El-Zahab B, Hamdan S, Karam T, Haber L, Li M, et al. Fluorescence, phosphorescence, and chemiluminescence. Analytical Chemistry. 2015;**88**:170-202. DOI: 10.1021/acs.analchem.5b04109

[5] Lagorio MG, Cordon GB, Iriel A. Reviewing the relevance of fluorescence in biological systems. Photochemical and Photobiological Sciences. 2015;**14**(9): 1538-1559. DOI: 10.1039/c5pp00122f

[6] Shimomura O, Johnson FH, Saiga Y. Extraction, purification and properties of Aequorin, a bioluminescent protein from the luminous hydromedusan, Aequorea. Journal of Cellular Physiology. 1962;**59**(3):223-239

[7] Tsien RY. Green fluorescent protein. Annual Review of Biochemistry. 1998;**67**:509-544. DOI: 10.1146/annurev.biochem.67.1.509

[8] Shagin DA, Barsova EV, Yanushevich YG, Fradkov AF, Lukyanov KA, Labas YA, et al. GFP-like proteins as ubiquitous metazoan superfamily: Evolution of functional features and structural complexity. Molecular Biology and Evolution. 2004;**21**: 841-850. DOI: 10.1093/molbev/msh079

[9] Goodwin RH. Fluorescent substances in plants. Annual Review of Plant Physiology. 1953;**4**:283-304

[10] Sparks JS, Schelly RC, Smith WL, Davis MP, Tchernov D, Pieribone VA, et al. The covert world of fish biofluorescence: A phylogenetically widespread and phenotypically variable phenomenon. PLoS One. 2014;**9**(1):e83259. DOI: 10.1371/journal.pone.0083259

[11] Gruber DF, Sparks JS. First observation of fluorescence in marine turtles. American Museum Novitates. 2015;**2015**(3845):1-9. DOI: 10.1206/3845.1

[12] Pogoda P. The fluorescence component in guppy colouration and how it relates to the light environment [PhD thesis]. Stuttgart: Stuttgart State Museum of Natural History; 2016. DOI: 10.13140/RG.2.2.27673.06249

[13] Taboada C, Brunetti AE, Pedron FN, Neto FC, Estrin DA, Bari SE, et al. Naturally occurring fluorescence in frogs. Proceedings of the National Academy of Sciences. 2017;**114**(14):3672-3677. DOI: 10.1073/pnas.1701053114

[14] Forge TA, Macguidwin AE. Nematode autofluorescence and its use as an indicator of viability. Journal of Nematology. 1989;**21**(3):399-403

[15] Viegas MS, Martins TC, Seco F, Carmo AD. An improved and cost-effective methodology for the reduction of autofluorescence in direct immunofluorescence studies in formalin-fixed paraffin-embedded tissues. European Journal of Histochemistry. 2007;**51**(1):59-66

[16] Winsor L. Studies on systematics and biogeography of terrestrial flatworms (Platyhelminthes: Tricladida: Terricola) of the Australia region [PhD thesis]. Townsville: James Cook University; 2003

[17] Winsor L. Taxonomic studies on free-living flatworms (Turbellaria:

Platyhelminthes) of the Australia zoogeographic region [Master's thesis]. Townsville: James Cook University; 1990

[18] Wei L, Li BY, Du HL, Zhou X, Gong W, Zhuge HX. Study on autofluorescence and confocal λ scanning characteristics of adult *Schistosoma japonicum*. Chinese Journal of Zoonoses. 2009;**25**(4):326-329

[19] Copeland J, Daston MM. Bioluminescence in the terrestrial snail *Quantula* (*Dyakia*) *striata*. Malacologia. 1988;**30**(1-2):317-324

[20] Isobe M, Uyakul D, Goto T, Counsilman J. *Dyakia* bioluminescence—1. Bioluminescence and fluorescence spectra of the land snail, *D. striata*. Journal of Bioluminescence and Chemiluminescence. 1988;**2**:73-79. DOI: 10.1002/bio.1170020204

[21] Isobe M, Uyakul D, Sigurdsson JB, Goto T, Lam TJ. Fluorescent substance in the luminous land snail, *Dyakia striata*. Agricultural and Biological Chemistry. 1991;**55**(8):1947-1951. DOI: 10.1080/00021369.1991.10870892

[22] Savazzi E, Sasaki T. Observations on land-snail shells in near-ultraviolet, visible and near-infrared radiation. Journal of Molluscan Studies. 2013;**79**(2):95-111. DOI: 10.1093/mollus/eys039

[23] Kirk RL, Main AR, Beyer FG. The use of paper partition chromatography for taxonomic studies of land snails. Biochemical Journal. 1954;**57**(3):440-442

[24] Seki K, Wiwegweaw A, Asami T. Fluorescent pigment distinguishes between sibling snail species. Zoological Science. 2008;**25**(12):1212-1219. DOI: 10.2108/zsj.25.1212

[25] Friend H. Luminous earthworms. Nature. 1893;**47**:462-463

[26] Harvey EN. Bioluminescence. New York: Academic Press Inc; 1952. p. 649

[27] Johnson FH, Shimomura O, Haneda Y. A note on the large luminescent earthworm, *Octochaetus multiporus*, of New Zealand. In: Johnson FH, Haneda Y, editors. Bioluminescence in Progress. Princeton: Princeton University Press; 1966. pp. 385-390

[28] Wampler JE, Jamieson BGM. Earthworm bioluminescence: Comparative physiology and biochemistry. Comparative Biochemistry and Physiology. 1979;**66b**:43-50

[29] Wampler JE. The bioluminescence system of *Microscolex phosphoreus* and its similarities to those of other bioluminescent earthworms (Oligochaeta). Comparative Biochemistry and Physiology Part A: Physiology. 1982;**71**(4):599-604. DOI: 10.1016/0300-9629(82)90209-2

[30] Rodionova NS, Rota E, Tsarkova AS, Petushkov VN. Progress in the study of bioluminescent earthworms. Photochemistry and Photobiology. 2017;**39**(2):416-428. DOI: 10.1111/php.12709

[31] Verdes A, Gruber DF. Glowing worms: Biological, chemical, and functional diversity of bioluminescent annelids. Integrative Comparative Biology. 2017;**57**(1):18-32. DOI: 10.1093/icb/icx017

[32] Wenig K, Kubista V. The presence of riboflavin in the luminous material of the earthworm *Eisenia submontana*. Experimentia. 1949;**5**:73

[33] Płytycz B, Homa J, Kozioł B, Rozanowska M, John Morgan A. Riboflavin content in autofluorescent earthworm coelomocytes is species-specific. Folia Histochemica et Cytobiologica. 2006;**44**:275-280. DOI: 10.5603/4560

[34] Pes O, Midlik A, Schlaghamersky J, Zitnand M, Taborskyand P. A study on bioluminescence and photoluminescence in the earthworm *Eisenia lucens*. Photochemical and Photobiological Sciences. 2016;**15**: 175-180. DOI: 10.1039/C5PP00412H

[35] Heredia RB, Dueñas S, Castillo L, Ventura JJ, Silva Briano M, Posadas del Rio F, et al. Autofluorescence as a tool to study mucus secretion in *Eisenia foetida*. Comparative Biochemistry and Physiology, Part A: Physiology. 2008;**151**(3):407-414. DOI: 10.1016/j.cbpa.2007.01.726

[36] Christyraj S, Retnaraj Samuel JR, Subramanian E, Vedha B, Edith Arul Jane A, Amutha K, et al. Autofluorescence in BrdU-positive cells and augmentation of regeneration kinetics by riboflavin. Stem Cells and Development. 2011;**21**:2071-2083. DOI: 10.1089/scd.2011.0485

[37] Albani JR, Sylvain D, Grumiaux F, Leprêtre A. Fluorescence fingerprints of *Eisenia fetida* and *Eisenia andrei*. Photochemistry and Photobiology. 2004;**78**:599-602. DOI: 10.1562/0031-8655(2003)0780599FFOE FA2.0.CO2

[38] Rorat A, Kachamakova-Trojanowska N, Jozkowicz A, Kruk J, Cocquerelle C, Vandenbulcke F, et al. Coelomocyte-derived fluorescence and DNA markers of composting earthworm species. Journal of Experimental Zoology. Part A, Ecological Genetics and Physiology. 2014;**321**(1):28-40. DOI: 10.1002/jez.1834

[39] Coburn C, Allman E, Mahanti P, Benedetto A, Cabreiro F, et al. Anthranilate fluorescence marks a calcium-propagated necrotic wave that promotes organismal death in *C. elegans*. PLoS Biology. 2013;**11**(7):e1001613. DOI: 10.1371/journal.pbio.1001613

[40] Robson EA. The cuticle of *Peripatopsis moseleyi*. Journal of Cell Science. 1964;**281-299**:s3-s105

[41] Lawrence RF. Fluorescence in Arthropoda. Journal of the Entomological Society of South Africa. 1954;**17**(2):167-170

[42] Shade DA. Morphological observations of fluorescence in different tick species (Family: Ixodidae). Acarologia. 2018;**58**:989-994. DOI: 10.24349/acarologia/20184305

[43] Rubin M, Lamsdell JC, Prendini L, Hopkins MJ. Exocuticular hyaline layer of sea scorpions and horseshoe crabs suggests cuticular fluorescence is plesiomorphic in chelicerates. Journal of Zoology. 2017;**303**:245-253. DOI: 10.1111/jzo.12493

[44] Filshie BK, Hadley NF. Fine structure of the cuticle of the desert scorpion, *Hadrurus arizonensis*. Tissue & Cell. 1979;**11**:249-262

[45] Andrews K, Reed SM, Masta SE. Spiders fluoresce variably across many taxa. Biology Letters. 2007;**3**(3):265-267. DOI: 10.1098/rsbl.2007.0016

[46] Marek P. Ultraviolet-induced fluorescent imaging for millipede taxonomy. Research Ideas and Outcomes. 2017;**3**:e14850. DOI: 10.3897/rio.3.e14850

[47] Marek PE, Moore W. Discovery of a glowing millipede in California and the gradual evolution of bioluminescence in Diplopoda. Proceedings of the National Academy of Sciences. 2015;**112**(20):6419-6424. DOI: 10.1073/pnas.1500014112

[48] Welch VL, Van Hooijdonk E, Intrater N, Vigneron JP. Fluorescence in insects. In: Proceedings SPIE—The International Society for Optical Engineering, The Nature of Light: Light in Nature IV; 848004. Vol. 8480. 2012. DOI: 10.1117/12.929547

[49] Wiesenborn WD. UV-excited fluorescence on riparian insects except Hymenoptera is associated with nitrogen content. Psyche. 2011;**875260**:1-6. DOI: 10.1155/2011/875250

[50] Frost LM, Butler DR, XO'Dell B, Fet V. A coumarin as a fluorescent compound in scorpion cuticle. In: Fet V, Selden PA, editors. Scorpions 2001: In Memoriam Gary A. Polis. Buckinghamshire: British Arachnological Society; 2001. pp. 365-368

[51] Gaffin DD, Bumm LA, Taylor MS, Popokina NV, Mann S. Scorpion fluorescence and reaction to light. Animal Behaviour. 2012;**83**(2):429-436. DOI: 10.1016/j.anbehav.2011.11.014

[52] Lourenço WR. Fluorescence in scorpions under UV light; can chaerilids be a possible exception? Comptes Rendus Biologies. 2012;**335**(12):731-734. DOI: 10.1016/j.crvi.2012.11.001

[53] Lourenço WR, Kovoor J, Munoz-Cuevas A. Observations on spiders in ultraviolet light. Tub-dokumentation Kongresse und Tagungen. Berlin. 1988;**38**:342-349

[54] Gould SJ, Lewontin RC. The spandrels of San Marco and the Panglossian Paradigm: A critique of the adaptationist program. Proceedings of the Royal Society London B: Biological Sciences. 1979;**205**:581-598. DOI: 10.1098/rspb.1979.0086

[55] Allmeling C, Jokuszies A, Reimers K, Kall S, Vogt PM. Use of spider silk fibres as an innovative material in a biocompatible artificial nerve conduit. Journal of Cellular and Molecular Medicine. 2007;**10**(3):770-777. DOI: 10.1111/j.1582-4934.2006.tb00436.x

[56] Heiling AM, Cheng K, Chittka L, Goeth A, Herberstein M. The role of UV in crab spider signals: Effects on perception by prey and predators. The Journal of Experimental Biology. 2005;**208**:3925-3931. DOI: 10.1242/jeb.01861

[57] Brandt EE, Masta SE. Females are the brighter sex: Differences in external fluorescence across sexes and life stages of a crab spider. PLoS One. 2017;**12**(5):e0175667. DOI: 10.1371/journal.pone.0175667

[58] Li J, Zhang Z, Liu F, Liu Q, Gan W, Chen J, et al. UVB-based mate-choice cues used by females of the jumping spider *Phintella vittata*. Current Biology. 2008;**18**(9):699-703. DOI: 10.1016/j.cub.2008.04.020

[59] Lim MLM, Land MFL, Li D. Sex-specific UV and fluorescence signals in jumping spiders. Science. 2007;**315**: 481-481. DOI: 10.1126/science.1134254

[60] Hsiung BK, Blackledge T, Skawkey M. Structural color and its interaction with other color-producing elements: Perspectives from spiders. In: Liang R, Shaw JA, editors. Proceedings of SPIE—The International Society for Optical Engineering, The Nature of Light: Light in Nature. Vol. 91870B. 2014. pp. 1-20. DOI: 10.1117/12.2060831

[61] Salih A, Larkum A, Cox G, Kühl M, Hough-Guldberg O. Fluorescent pigments in corals are photoprotective. Nature. 2000;**408**:850-853. DOI: 10.1038/35048564

[62] Croucher PJP, Brewer MS, Winchell CJ, Oxford GS, Gillespie RG. De novo characterization of the gene-rich transcriptomes of two color-polymorphic spiders, *Theridion grallator* and *T. californicum* (Araneae: Theridiidae), with special reference to pigment genes. BMC Genomics. 2013;**14**(1):862. DOI: 10.1186/1471-2164-14-862

[63] Nappi AJ, Christensen BM. Melanogenesis and associated cytotoxic reactions: Applications to insect innate

immunity. Insect Biochemistry and Molecular Biology. 2005;**35**(5):443-459. DOI: 10.1016/j.ibmb.2005.01.014

[64] Hsiung BK, Blackledge T, Skawkey M. Spiders do have melanin after all. Journal of Experimental Biology. 2015;**218**:3632-3635. DOI: 10.1242/jeb.128801

[65] Kuse K, Yanagi M, Tanaka E, Tani N, Nishikawa T. Identification of a fluorescent compound in the cuticle of the train millipede *Parafontaria laminate armigera*. Bioscience, Biotechnology, Biochemistry. 2010;**74**(11):2307-2309. DOI: 10.1271/bbb.100171

[66] Anderson JM. Biochemistry of centipede bioluminescence. Photochemistry and Photobiology. 1980;**31**(2):179-181. DOI: 10.1111/j.1751-1097.1980.tb03701.x

[67] Rosenberg J, Meyer-Rochow VB. Luminescent myriapoda: A brief review. In: Meyer-Rochow VB, editor. Bioluminescence in Focus—A Collection of Illuminating Essays. Trivandrum: Research Signpost; 2009. pp. 139-146

[68] Oba Y, Stevani CV, Oliveira AG, Tsarkova AS, Chepurnykh TV, Yampolsky IV. Selected least studied but not forgotten bioluminescent systems. Photochemistry and Photobiology. 2016;**93**(2):405-415. DOI: 10.1111/php.12704

[69] Kuse M, Kanakubo A, Suwan S, Koga K, Isobe M, Shimomura O. 7,8-Dihydropterin-6-carboxylic acid as light emitter of luminous millipede, *Luminodesmus sequoiae*. Bioorganic and Medicinal Chemistry Letters. 2001;**11**:1037-1040. DOI: 10.1016/S0960-894X(01)00122-6

[70] Marek P, Papaj D, Yeager J, Molina S, Moore W. Bioluminescent aposematism in millipedes. Current Biology. 2011;**21**(18):R680-R681. DOI: 10.1016/j.cub.2011.08.012

[71] Lloyd JE. Bioluminescence and communication in insects. Annual Review of Entomology. 1983;**28**:131-160. DOI: 10.1146/annurev.en.28.010183.001023

[72] Zompro O, Fritzsche I. *Lucihormetica* n. gen. n. sp., the first record of luminescence in an orthopteroid insect (Dictyoptera: Blaberidae: Blaberinae: Brachycolini). Amazoniana. 1999;**15**:211-219

[73] Merritt DJ. Standards of evidence for bioluminescence in cockroaches. Die Naturwissenschaften. 2013;**100**(7):697-698. DOI: 10.1007/s00114-013-1067-9

[74] Beckert J, Greven H, Lunau K. UV-reflection and autofluorescence of the pronotal spots in the glowspot Cockroach *Lucihormetica verrucosa* (Brunner von Wattenwyl, 1865) (Blattodea: Blaberidae) are affected by carotenoid diet. Entomologie Heute. 2017;**29**:25-33

[75] Tokumoto H, Shimomura H, Hakamatsuka T, Ozeki Y, Goda Y. Fluorescence coupled with macro and microscopic examinations of morphological phenotype give key characteristics for identification of crude drugs derived from scorpions. Biological and Pharmaceutical Bulletin. 2018;**41**(4):510-523. DOI: 10.1248/bpb.b17-00801

[76] Brown DJ. Fused pyrimidines, part 3: Pteridines. In: Taylor E, editor. The Chemistry of Heterocyclic Compounds. Vol. 24. New York: John Wiley & Sons Inc.; 1988. 731 p

[77] Rawson GW. Study of fluorescent pigments in Lepidoptera by means of paper partition chromatography. Journal of Lepidopterists' Society. 1968;**22**(1):27-40

[78] Zill S, Frazier SF, Neff D, Quimby L, Carney M, Dicaprio R, et al. Three-dimensional graphic reconstruction of the insect exoskeleton through confocal imaging of endogenous fluorescence. Microscope Research and Technique. 2000;**48**:367-384. DOI: 10.1002/(SICI)1097-0029(20000315)48:6<367::AID-JEMT7>3.0.CO;2-Y

[79] Neff D, Frazier SF, Quimby L, Wang RT, Zill S. Identification of resilin in the leg of cockroach, *Periplaneta americana*: Confirmation by a simple method using pH dependence of UV fluorescence. Arthropoda Structure and Development. 2000;**29**(1):75-83. DOI: 10.1016/S1467-8039(00)00014-1

[80] Kishi Y, Matusuura S, Inoue S, Shimonura O, Goto T. Luciferin and luciopterin isolated from the Japanese firefly, *Luciola cruciata*. Tetrahedron Letters. 1968;**24**:2847-2850. DOI: 10.1016/S0040-4039(00)75644-9

[81] Donoughe S, Crall JD, Merz RA, Combes RA. Resilin in dragonfly and damselfly wings and its implications for wing flexibility. Journal of Morphology. 2011;**272**:1409. DOI: 10.1002/jmor.10992

[82] Philips LS. Fluorescence in the colors of certain Lepidoptera observed under ultraviolet light. Journal of the Lepidoperists' Society. 1959;**13**(2):73-77

[83] Guillermo-Ferreira R, Therézio EM, Gehlen MH, Bispo PC, Marletta A. The role of wing pigmentation, UV and fluorescence as signals in a neotropical damselfly. Journal of Insect Behavior. 2014;**27**(1):67-80. DOI: 10.1007/s10905-013-9406-4

[84] van Hooijdonk E, Barthou C, Vigneron JP Berthier S. Yellow structurally modified fluorescence in the longhorn beetles *Celosterna pollinosa sulfurea* and *Phosphorus virescens* (Cerambycidae). Journal of Luminescence. 2013;**136**:313. DOI: 10.1016/j.jlumin.2012.12.022

[85] Marshall J, Johnsen S. Fluorescence as a means of colour signal enhancement. Philosophical Transactions of the Royal Society B. 2016;**372**:20160335. DOI: 10.1098/rstb.2016.0335

[86] Völker O. Über fluoreszierende, gelbe Federpigmente bei Papageien, eine neue Klasse von Federfarbstoffen. Journal of Ornithology. 1937;**85**:136-146. DOI: 10.1007/BF01905492

[87] Arnold KE, Owens IPF, Marshall NJ. Fluorescent signaling in parrots. Science. 2002;**295**(5552):92. DOI: 10.1126/science.295.5552.92

[88] McGraw KJ, Toomey MB, Nolan PM, Morehouse NA, Massaro M, Jouventin P. A description of unique fluorescent yellow pigments in penguin feathers. Pigment Cell and Melanoma Research. 2007;**20**:301-304. DOI: 10.1111/j.1600-0749.2007.00386.x

[89] Prötzel D, Heß M, Scherz MD, Schwager M, Padje AV, Glaw F. Widespread bone-based fluorescence in chameleons. Scientific Reports. 2018;**8**:698. DOI: 10.1038/s41598-017-19070-7

[90] Kohler AM, Olson ER, Martin JG, Anich PS. Ultraviolet fluorescence discovered in New World flying squirrels (*Glaucomys*). Journal of Mammalogy. 2019;**100**(1). DOI: 10.1093/jmammal/gyy177

[91] Weagle G, Paterson PE, Kennedy J, Pottier R. The nature of the chromophore responsible for naturally occurring fluorescence in mouse skin. Journal of Photochemistry and Photobiology B: Biology. 1988;**2**:313-320. DOI: 10.1016/1011-1344(88)85051-6

[92] Parker AR. Fluorescence of yellow budgerigars. Science.

2002;**296**(5568):655. DOI: 10.1126/science.296.5568.655b

[93] Pearn SM, Bennett ATD, Cuthill IC. The role of ultraviolet–A reflectance and ultraviolet–A induced fluorescence in the appearance of budgerigar plumage: Insights from spectrofluorometry and reflectance spectrophotometry. Proceedings of the Royal Society of London, B. 2003;**270**:859-865. DOI: 10.1098/rspb.2002.2315

[94] Endler JA. The color of light in forests and its implications. Ecological Monographs. 1993;**36**:1-27

[95] Stuart-Fox D, Moussalli A, Whiting MJ. Natural selection on social signals: Signal efficacy and the evolution of chameleon display coloration. American Naturalist. 2007;**170**(6): 916-930. DOI: 10.1086/522835

[96] Dubois R. Les Elatérides Lumineux. Bulletin de la Société zoologique de France. 1896;**11**:1-275

[97] Harvey NE. The Nature of Animal Light. Philadelphia: J.B. Lippincott Company; 1920. 118 p

[98] Coblentz WW. Notiz über eine von der Feuerfliege herrührende fluoreszierende Substanz. Physikalische Zeitschrift. 1909;**10**:955-956

[99] Dubois R. La Vie et la lumière: Biophotogénèse ou production de la lumière par les êtres vivants; action de la lumière visible, des radiations ultra-violettes, infra-rouges, fluorescentes, des rayons X, du radium et des ondes hertziennes sur les animaux et sur les végétaux; photothérapie. Paris: F. Alcan; 1914. 338 p

[100] McDermott FA. Luciferesceine, the fluorescent material present in certain luminous insects. Journal of the American Chemical Society. 1911;**33**:410-416

[101] Strehler BL. The isolation and properties of firefly luciferesceine. Archives of Biochemistry and Biophysics. 1951;**32**(2):397-406. DOI: 10.1016/0003-9861(51)90289-5

[102] Bitler B, McElroy WD. The preparation and properties of crystalline firefly luciferin. Archives of Biochemistry and Biophysics. 1957;**72**:358-368

[103] Gandelman OA, Brovko LY, Ugarova NN, Chikishev AY, Shkurinov AP. Oxyluciferin fluorescence is a model of native bioluminescence in the firefly luciferin-luciferase system. Photochemistry and Photobiology B: Biology. 1993;**19**:187-191. DOI: 10.1016/1011-1344(93)87083-Y

[104] Shimomura O. Bioluminescence: Chemical Principles and Methods. Hackensack: World Scientific; 2006. DOI: 10.1142/9789812773647

[105] Naumov P, Ozawa Y, Ohkubo K, Fukuzumi S. Structure and spectroscopy of oxyluciferin, the light emitter of the firefly bioluminescence. Journal of the American Chemical Society. 2009;**131**(32):11590-11605. DOI: 10.1021/ja904309q

[106] Pinto da Silva L, Esteves da Silva JCG. Theoretical analysis of the color tuning mechanism of oxyluciferin and 5-hydroxyoxyluciferin. Computational and Theoretical Chemistry. 2012;**988**:56-62. DOI: 10.1016/j.comptc.2012.02.029

[107] Vieira J, Pinto da Silva L, Esteves da Silva JCG. Advances in the knowledge of light emission by firefly luciferin and oxyluciferin. Journal of Photochemistry and Photobiology B: Biology. 2012;**117**: 33-39. DOI: 10.1016/j.jphotobiol.2012.08.017

[108] Smalley KN, Tarwater DE, Davidson TL. Localization of fluorescent compounds in the firefly light organ. Journal of Histochemistry

& Cytochemistry. 1980;**28**(4):323-329. DOI: 10.1177/28.4.7373026

[109] Metcalf RL. The isolation of a red-fluorescent pigment, lampyrine, from the Lampyridae. Annals Entomological Society of America. 1943;**36**:37-40

[110] Wilkerson RC, Lloyd J. The application of paper chromatography of fluorescent compounds to the systematics of fireflies (Coleoptera, Lampyridae). The Coleopterists' Bulletin. 1975;**29**(4):339-348

[111] Sannasi A. Resilin in the lens cuticle of the firefly, *Photinus pyralis* Linnaeus. Experimentia. 1970;**26**(2):154-154. DOI: 10.1007/BF01895549

[112] Yiu V, Jeng ML. *Oculogryphus chenghoiyanae* sp. n. (Coleoptera, Lampyridae): A new ototretine firefly from Hong Kong with first descriptions of the female and bioluminescent behavior of the genus and ultraviolet-induced fluorescence in the family. Zookeys. 2018;**739**:5678. DOI: 10.3897/zookeys.739.21502

[113] Michels J, Appel E, Gorb SN. Functional diversity of resilin in Arthropoda. Beilstein Journal of Nanotechnology. 2016;**7**:1241-1259. DOI: 10.3762/bjnano.7.115

[114] Kawashima I. The lampyrid beetles of the genus *Stenocladius* (Coleoptera, Lampyridae) of the Ryukyu Islands, Southwest Japan, with descriptions of two new species. Elytra. 1999;**27**:141-158

[115] Chen TR. The Fireflies of Taiwan. Taipei City: Field Image Press; 2003. 255 p [In Chinese]

[116] Hall DW, Sander SE, Pallansch JC, Stanger-Hall KF. The evolution of adult light emission color in North American fireflies. Evolution: An International Journal of Organic Evolution. 2016;**70**(9):2033-2048. DOI: 10.1111/evo.13002

[117] Johnsen S, Kelber A, Warrant EJ, Sweeney AM, Widder EA, Lee RL Jr, et al. Crepuscular and nocturnal illumination and its effects on color perception by the nocturnal hawkmoth *Deilephila elpenor*. Journal of Experimental Biology. 2006;**209**: 789-800. DOI: 10.1242/jeb.02053

[118] De Cock R, Matthysen E. Glow-worm larvae bioluminescence (Coleoptera: Lampyridae) operates as an aposematic signal upon toads (*Bufo bufo*). Behavioral Ecology. 2003;**14**(1):103-108. DOI: 10.1093/beheco/14.1.103

[119] Briscoe AD, Chittka L. The evolution of color vision in insects. Annual Reviews of Entomology. 2001;**46**:471-510. DOI: 10.1146/annurev.ento.46.1.471

[120] Sander SE, Hall DW. Variation in opsin genes correlates with signaling ecology in North American fireflies. Molecular Ecology. 2015;**24**(18): 4679-4696. DOI: 10.1111/mec.13346

[121] Martin GJ, Lord NP, Branham MA, Bybee SM. Review of the firefly visual system (Coleoptera: Lampyridae) and evolution of the opsin genes underlying color vision. Organisms Diversity & Evolution. 2015;**15**(3):513-526. DOI: 10.1007/s13127-015-0212-z

[122] Lall A, Lord E, Trouth C. Vision in the firefly *Photuris lucicrescens* (Coleoptera: Lampyridae): Spectral sensitivity and selective adaptation in the compound eye. Journal of Comparative Physiology A, Sensory, Neural, and Behavioral Physiology. 1982;**147**:195-200. DOI: 10.1007/BF00609844

[123] Booth D, Stewart AJA, Osorio D. Colour vision in the glow-worm *Lampyris noctiluca* (L.) (Coleoptera: Lampyridae): evidence for a green-blue

chromatic mechanism. Journal of Experimental Biology. 2004;**207**: 2373-2378. DOI: 10.1242/jeb.01044

[124] Owens ACS, Meyer-Rochow VB, Yang EC. Short- and mid-wavelength artificial light influences the flash signals of *Aquatica ficta* fireflies (Coleoptera: Lampyridae). PLoS One. 2018;**13**(2):e0191576. DOI: 10.1371/journal.pone.0191576